Introduction

The Paul Scherrer Institute PSI has four particle accelerators, which are described in this booklet.

1. Proton Ring Cyclotron 590 MeV

The project for a powerful proton accelerator started at ETHZ, the Swiss Federal Institute of Technology in Zurich, under professor J.P.Blaser in 1962. From the physicist Paul Scherrer he took over a small project group and redirected them towards the plan for a so-called meson factory. Hans Willax, the leader of this project, came up with a brilliant idea: he split up the conventional magnet of a cyclotron into separate sector magnets, making room in between them for powerful RF cavities. The first protons from this novel ring cyclotron were extracted in January 1974. The design goal of a current of 100 µA was reached in 1976. From the beginning it was realized that the current limit of the ring cyclotron was much higher than the one from the 72 MeV injector cyclotron from the Philips company. Thus a new Injector II cyclotron was conceived, and started operation in 1984. To increase the beam current to new and unexplored limits, the RF cavities of the ring cyclotron were replaced with more powerful ones. In 2009 a record value of 2.4 mA beam current could be reached, a value 24 times higher than the already ambitious design goal. At 590 MeV the average beam power is thus 1.4 MW, a world record to this day. The high energy protons produce spallation neutrons for materials science as well as pions and muons in record quantities.

2. Compact superconducting Proton Cyclotron, 250 MeV, used for irradiation of tumours

Starting in 1984, protons from the original 72 MeV cyclotron were used to irradiate eye tumours. But to eliminate deep-seated tumours one needs higher energy protons. Starting in 1996 a very small fraction of the 590 MeV proton beam was first degraded to energies between 100 and 250 MeV and then directed towards Gantry 1. With the so-called 3-dimensional spot scanning technique, invented by the physicist Eros Pedroni, tumours with complicated shapes can be irradiated. With steering magnets the proton beam can be deflected in the horizontal and vertical direction. In the third dimension - the direction of the beam - the range of the stopped beam can be changed by scanning the proton energy.

The increased demand of this successful method required an independent accelerator. The choice fell on a compact superconducting cyclotron from the German company ACCEL, based on ideas from the physicist Henry Blosser from Michigan State University. Irradition with protons has one big advantage over X-rays: it gives lower doses to the healthy tissue surrounding the tumour. This is very important when brain tumours or tumours close to the spinal cord are irradiated. It is especially crucial for the treatment of children. To treat more of these critical cases, and to further develop the scanning technology, Gantry 2 was built and came into operation in 2013. The latest treatment device (Gantry 3) was installed in a research collaboration with the company Varian Medical Systems, and started clinical operation in 2018. To date, over 8'000 patients have benefited from PSI's longstanding expertise with proton therapy.

3. Swiss Light Source SLS

Electrons, circulating in a storage ring of 280m circumference, produce intense X-ray beams, as thin as a hair, for materials science. The electrons are accelerated in a linear accelerator and in a booster synchrotron to the final energy of 2.4 GeV, before being injected into a storage ring. There the electrons make about one million revolutions each second. About 0.5% of the circulating electrons are lost in the next 2.5 minutes due to collisions with residual molecules, or with other electrons in the beam. These lost electrons are replaced by turning on the booster again for about two seconds. This so called top-up-operation was pioneered at PSI. It keeps the intensity of the circulating beam practically constant at a level between 400mA and 402mA. The SLS project was granted by the Swiss Parliament in 1997, thanks to the efforts of former PSI directors Meinrad Eberle and Wilfred Hirt. Starting in 2000 with four X-ray beamlines, there are now 23 in operation, producing beams from infrared to hard X-rays around the clock. Especially intense are the X-rays produced by so-called undulators, installed in some straight sections of the storage ring. The bright beam of hard X-rays makes it possible to reconstruct microscopic objects, e.g. blood vessels, in three dimensions. The determination of the structure of protein molecules is vital for the development of new pharmaceuticals. The pharmaceutical industry in Basel was initially sceptical about the SLS, but later sponsored a special beamline for protein crystallograpy. Beam time on the SLS is in great demand from groups all over the world, and is heavily overbooked.

4. SwissFEL

With the intense and bright X-rays from the SLS facility it is possible to make pictures of very small objects, using 3-dimensional micro-tomography. However the X-ray pulses are not short enough to image moving molecules. On the other hand the new facility SwissFEL produces short enough flashes of X-rays, in the range of femtoseconds, to make this possible. We can thus observe "dancing molecules". Following chemical reactions in small time intervals opens a completely new field in chemistry.

The term FEL stands for "Free Electron Laser". The advantage over short pulses from conventional lasers is the ability to vary the wavelength over a wide range between 0.1nm and 7nm. At the front end of SwissFEL an electron gun produces extremely short electron pulses, a few femtoseconds long. The electrons are then accelerated in a linear accelerator over a length of about 500m to an energy up to 6 GeV. Passing through so-called undulators, the electrons are forced by permanent magnets onto a slalom-like course and produce the desired very short pulses of X-rays.

The SwissFEL was inaugurated at the end of 2016 by Joel Mesot, the director of PSI. First experiments started in november 2017.

 Werner Joho was born 1938 in Baden, Switzerland, and now lives with his wife Rosa in Wuerenlingen, close to Baden. Like Einstein, he attended the cantonal high school in Aarau, and went on to study physics at ETHZ, the Swiss Federal Institute of Technology in Zurich. There he met Prof. Eduard Stiefel, an outstanding teacher, from whom he acquired a life-long enthusiasm for "Applied Mathematics". Then followed a fortunate turning-point when he joined the cyclotron group of Prof. J.P. Blaser, which was working on a project for a new high intensity proton cyclotron.

Calculating the orbits in this accelerator became his main duty, initially in 1962 at CERN, the European Research Center in Geneva. He also spent some time as a student at Michigan State University, working part-time on its cyclotron project. From 1964 onwards he was leader of the beam dynamics group at the Swiss Research Institute SIN (later PSI) in Villigen. He helped to optimize the performance of the Proton Ring Cyclotron and the Injector II cyclotron. This accelerator facility was pushed later to worlwide record intensities. In 1970 he graduated Ph.D. from ETH Zurich, under the supervision of J.P.Blaser. The topic of the thesis was the extraction of the proton beam from the ring cyclotron. From 1971 to 1973 he took a sabbatical leave at the TRIUMF institute of the University of British Columbia in Vancouver.

In 1990 the author decided to switch from accelerating protons to accelerating electrons. He spent some time at the project of the Advanced Light Source ALS at the University of Berkeley in California. After his return to PSI he helped to initiate the project for the Swiss Light Source SLS, where he was responsible for the construction of the booster synchrotron.

As a hobby, the author and two ETH colleagues, Jerry Tripard and Gerhard Rudolf, developed a computer chess program called CHARLY. On October 7th 1968 this program played for the first time a live game over the Atlantic (via ham radio) with a corresponding computer program MAC HACK VI. This program was developed by the department of Artificial Intelligence under Richard Greenblatt at MIT in Boston. The game lasted 3½ hours and the American program (using the "Russian defense" !) won after 41 moves. The moves of this historic game can be found e.g. with Google: "chess programming Joho". Over the years the author has collected many mathematical curiosities, a selection of which he published in 2018 in a booklet called "Winning Bets with Math", by the company "BoD – Books on Demand" (ISBN 978-3-7460-5623-4).

The author can be contacted at: werner.joho@bluewin.ch

In his free time Werner Joho has enjoyed many sports activities, like orienteering competitions in forests, skiing, tennis and windsurfing. For the past 30 years he has been an enthusiastic golfer.

The Paul Scherrer Institute PSI in Villigen/Wuerenlingen is the largest National Research Institute in Switzerland. It is sponsored by the Swiss Government and currently has 2'100 employees (750 externally financed). The dominant research areas are: matter and materials, human health, energy and environment. PSI has four state-of-the-art particle accelerators, two proton cyclotrons and two electron accelerators. They are used by scientists from all over the world.

The author has spent his whole career contributing to the concepts and operation for some of these accelerators. The following illustrations are mainly from talks given by him over many years. Several illustrations are freely borrowed from scientific publications by researchers using the accelerator facilities of PSI.
Some material was graciously supplied by the staff of the visitor center "Forum", which organizes guided tours for about 15'000 visitors every year!
Special thanks go to John Crawford, David Meer, Lenny Rivkin for helpful comments, and to Bruno Fricker for help with the cover.

© 2019 Werner Joho

3.Edition

Production and Publishing House:

BoD – Books on Demand,

Norderstedt, Germany

ISBN 9783-7528-47116

PSI has 4 Top-class Accelerators!

- Ring-Cyclotron : 590 MeV Protons (1974)

  => Neutrons, Pions, Muons

- Compact-Cyclotron COMET: 250 MeV Protons (2007)

  => Cancer Therapy

- SLS: Storage Ring , 2.4 GeV Electrons (2000)

  => X-rays

- SwissFEL, Electron Linac , 6 GeV (2016)

  => Femtosecond X-ray Flashes

3 Probes for Material Research

only at PSI and Rutherford Lab

Photons (SLS, SwissFEL) ⟹ Electron Cloud

Neutrons (SINQ) ⟹ Atomic Nuclei

Muons (μSR) ⟹ internal magnetic Fields in Crystals

Research Facilities in Europe

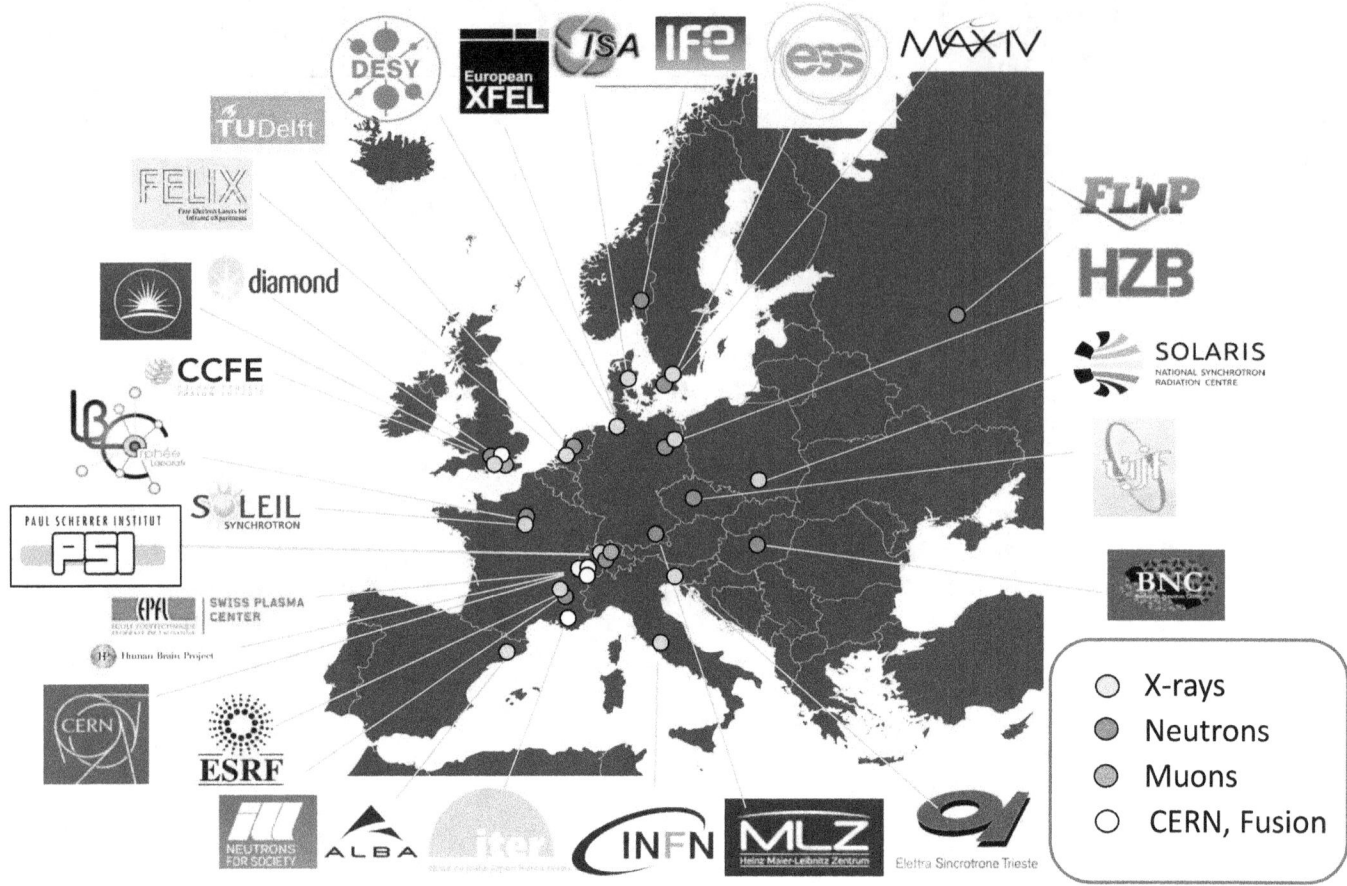

Ring Cyclotron 590 MeV Protons

2.4 mA, 1.4 MW average Beam Power (World Record!)

most intensive Muon Beams
$5 \cdot 10^8\ \mu^+/s$, $10^8\ \mu^-/s$

Spallation-Neutron-Source 10^{14} n/s
↓
equivalent to medium Flux Reactor (but without Uranium!)

Swiss Light Source (SLS)
2.4 GeV Electron Storage Ring

- constant beam current (400-402 mA)

  due to **top-up** injection every 2½ min.

- extremely stable photon beams

  due to „**fast orbit feedback**" (< 0.2 µm)

Nobel Prize in Chemistry 2009 !

V. Ramakrishnan won Nobel prize in chemistry

He was a user of a protein crystallography beam line

at the SLS at PSI

Investigation of Ribosomes with x-ray diffraction

Superconducting Cyclotron 250 MeV Protons for Cancer Therapy

70 MeV → Eye Tumours

70–250 MeV →
- deep seated Tumours
- three rotating Gantries
- 3D-Spot Scanning

Comet Cyclotron

Radiation Therapy with 250 MeV Protons
superconducting Cyclotron:
Magnet, 3m Ø
Collaboration: ACCEL & PSI

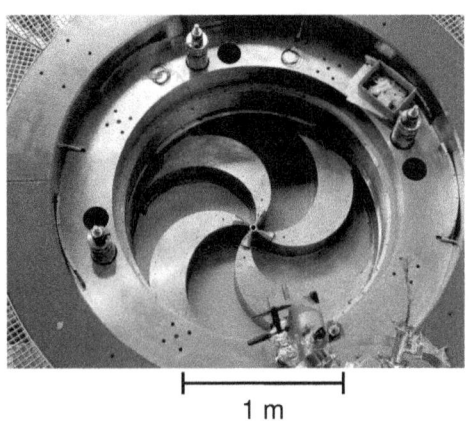

1 m

The spiral structure is responsible for the vertical beam focusing

The Medical Annex PROSCAN for Proton Therapy

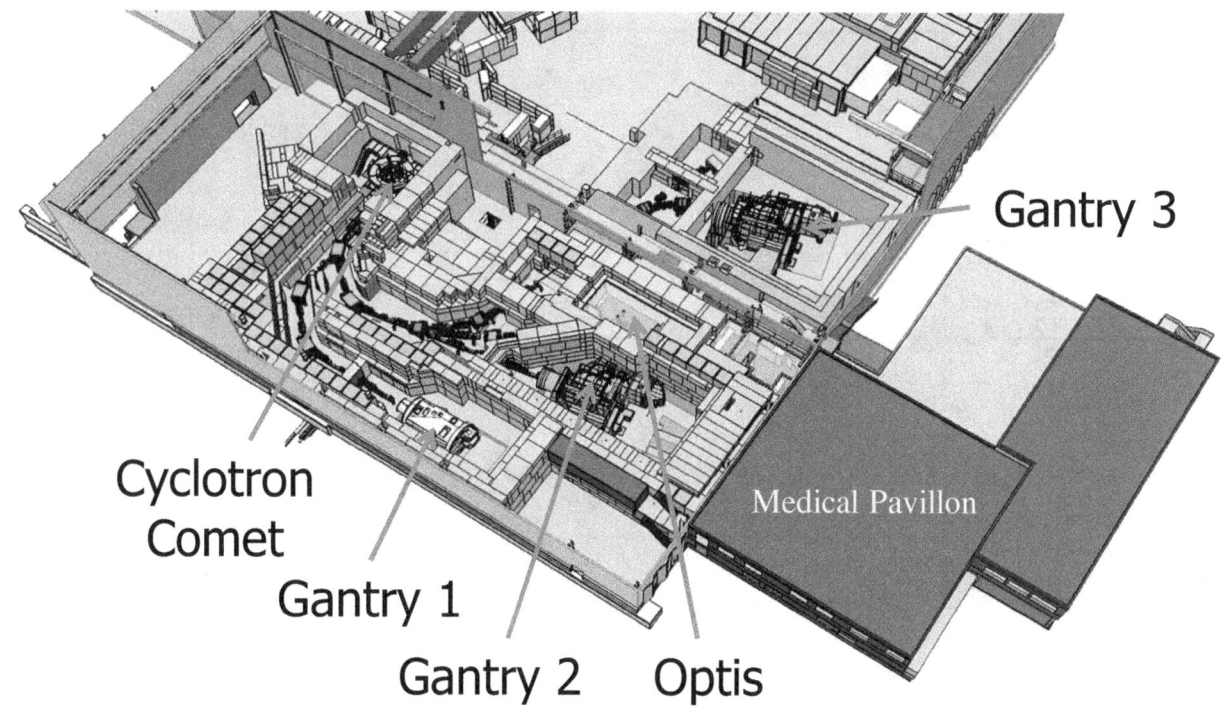

Cyclotron Comet
Gantry 1
Gantry 2
Optis
Gantry 3
Medical Pavillon

OPTIS, Eye Tumour Therapy with Protons

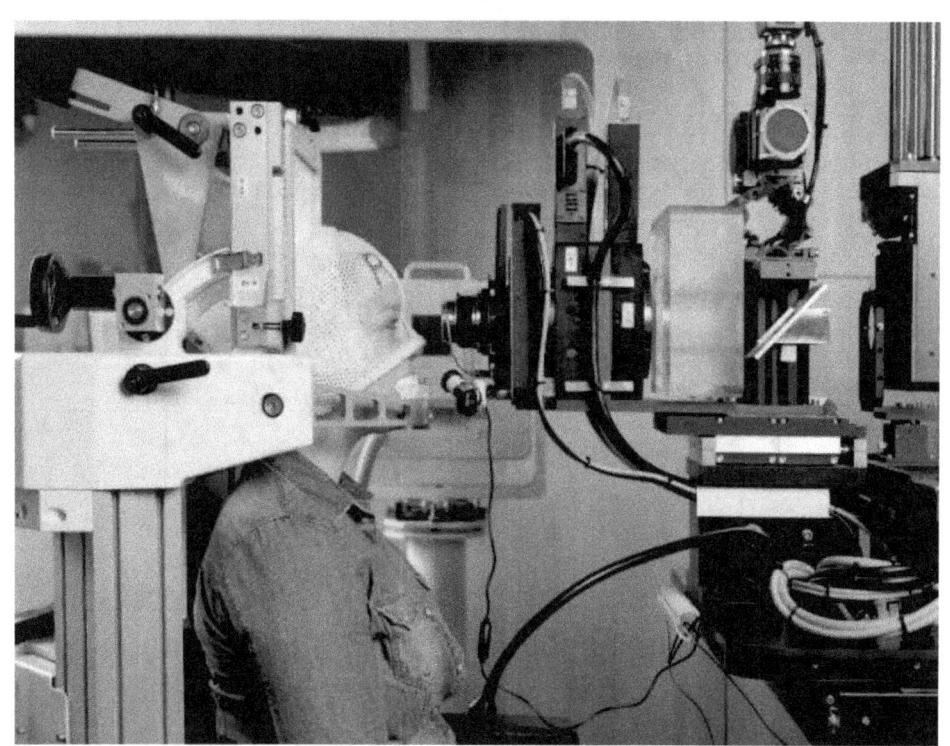

1984-2018

> 7'000 Patients

Tumour Control:

> 98%

(Collaboration with Hospital Lausanne)

Irradiation on 4 Days

Why Protons against Tumours?

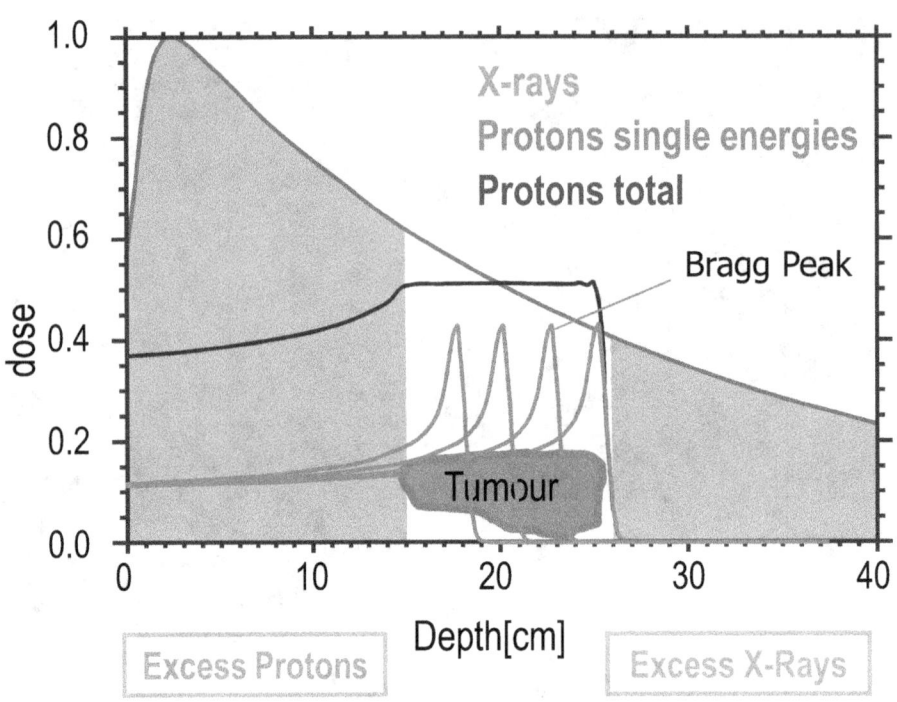

SPOT SCANNING

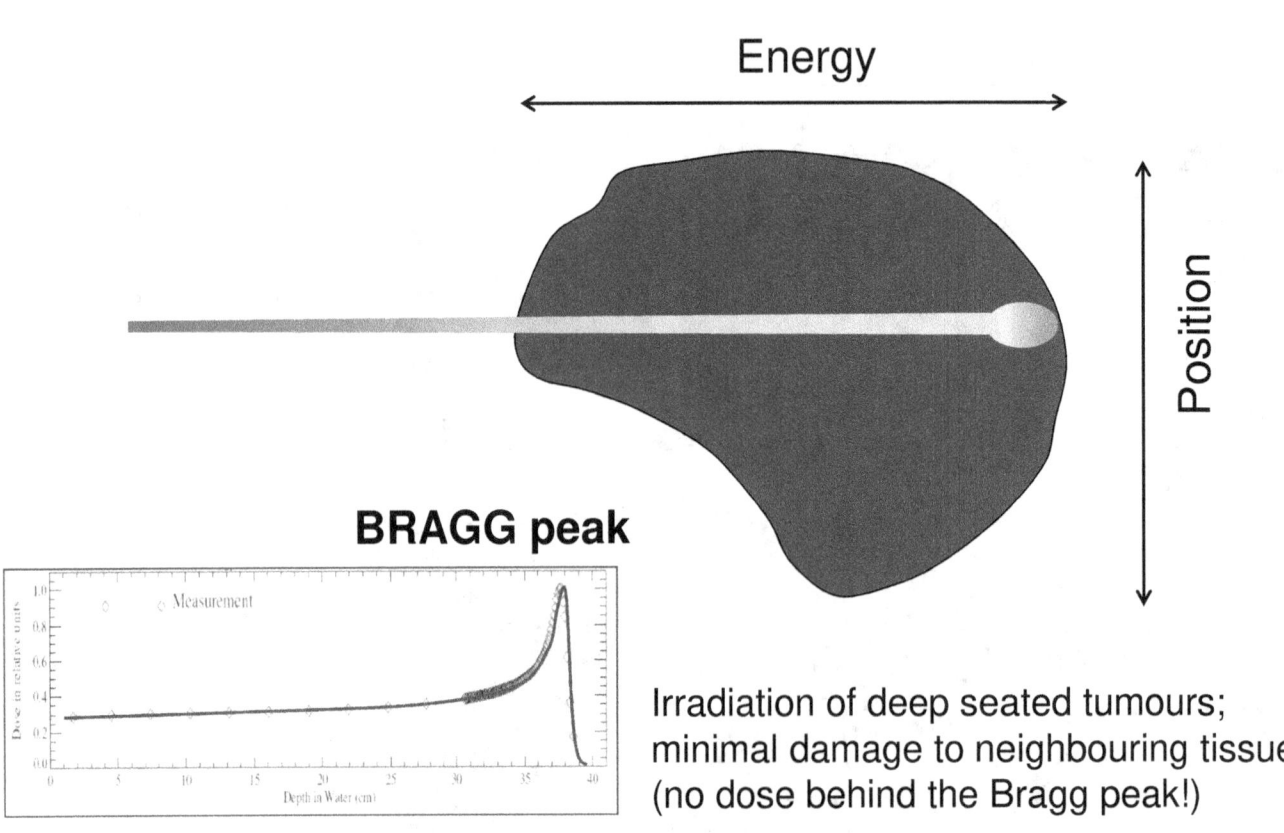

BRAGG peak

Irradiation of deep seated tumours; minimal damage to neighbouring tissues (no dose behind the Bragg peak!)

Brain Tumour

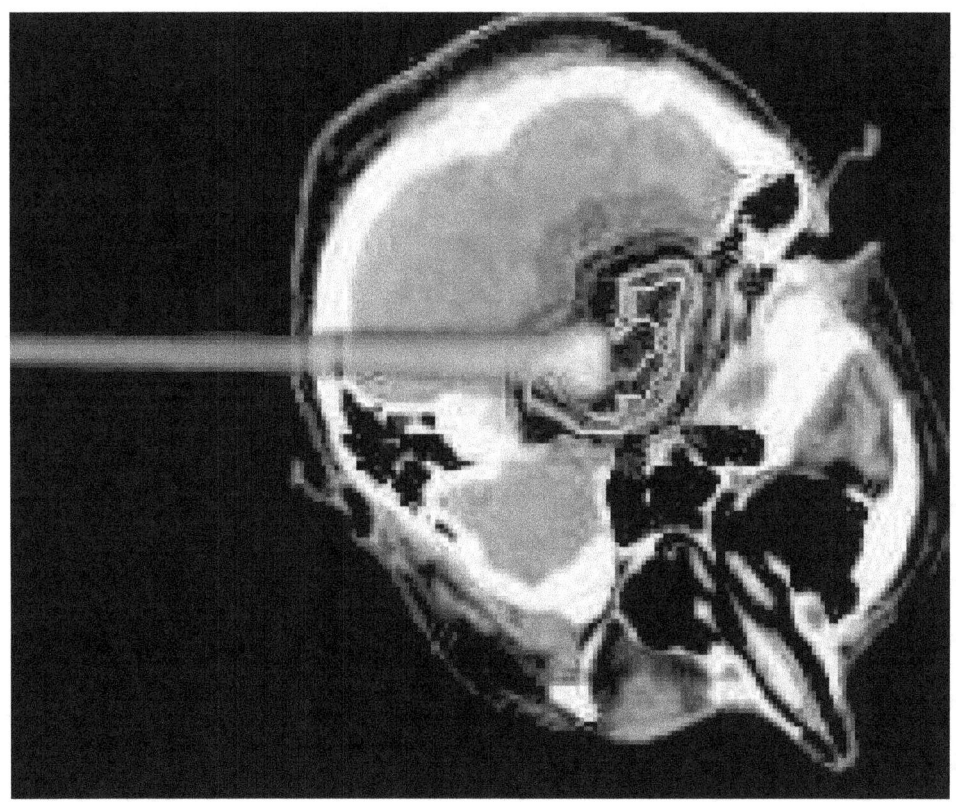

Irradiation with Protons by Spot-Scanning
(E.Pedroni, PSI)

Proton Therapy Gantry 1

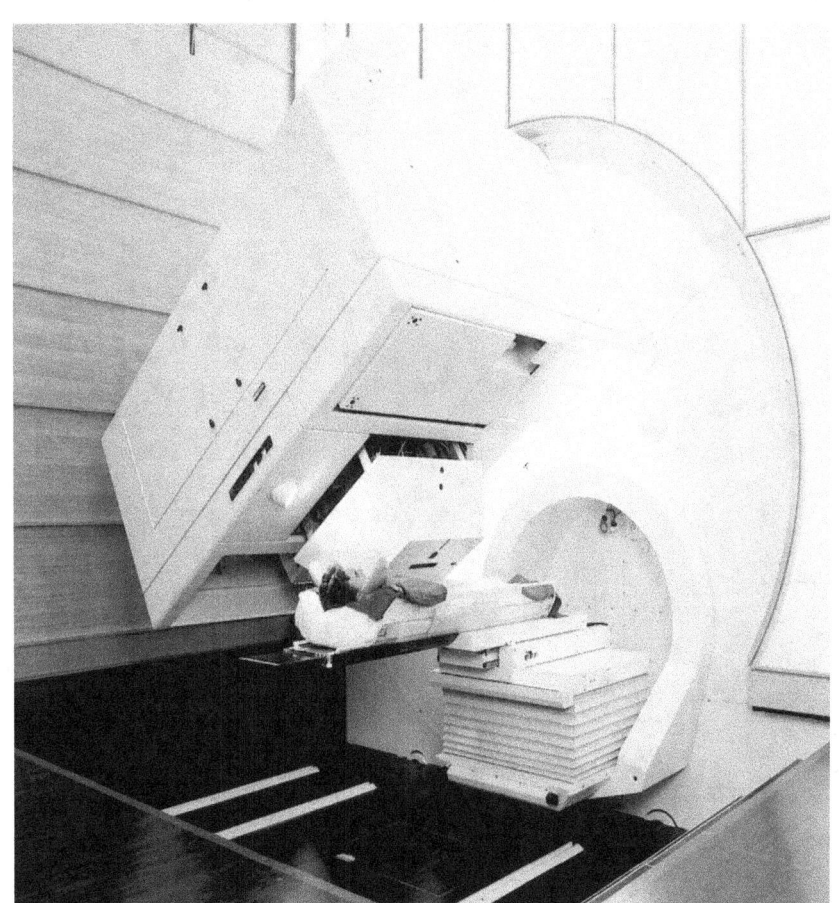

Irradiation of a Tumour from different Directions
$\Rightarrow$ minimal Dose at Surface

Proton Therapy Gantry 2

Irradiation of Tumours with Gantry 2

(built to treat tumours that move due to respiration)

Proton Therapy with Gantry 3

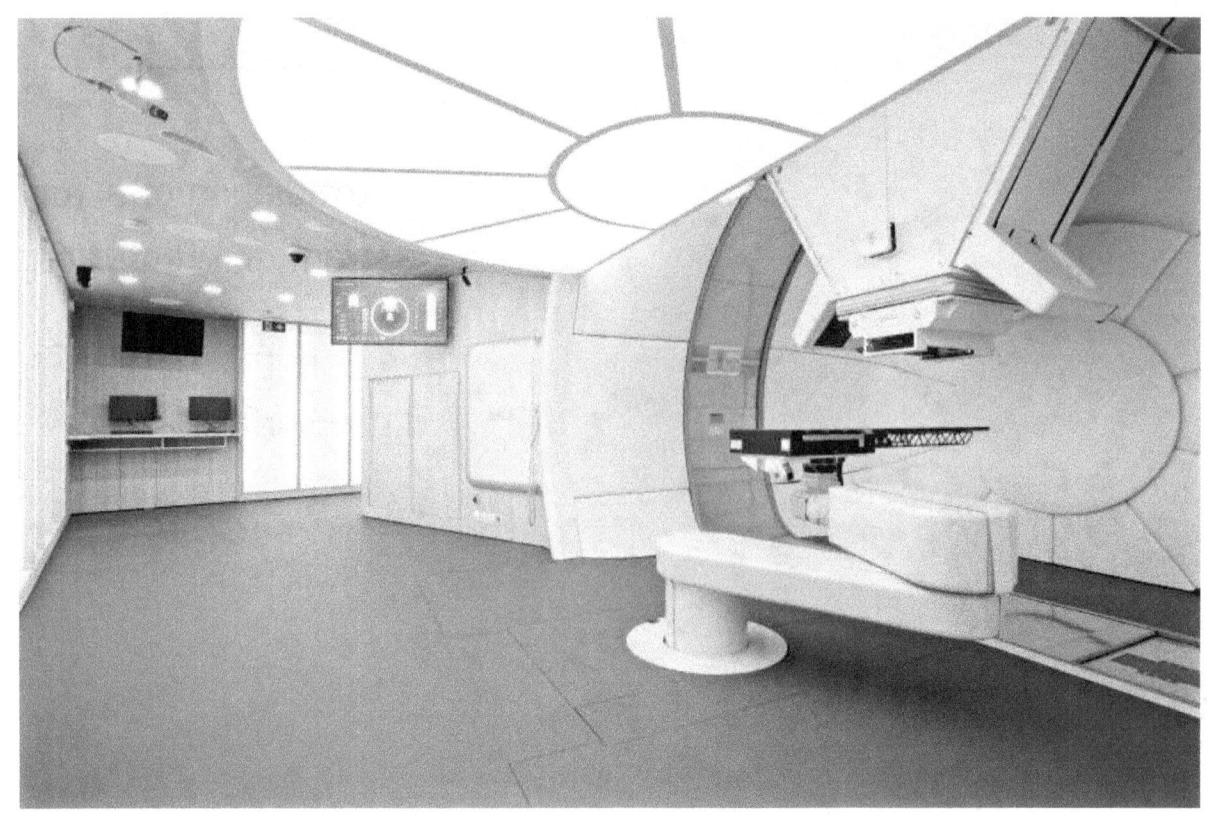

The first Cyclotron 1931

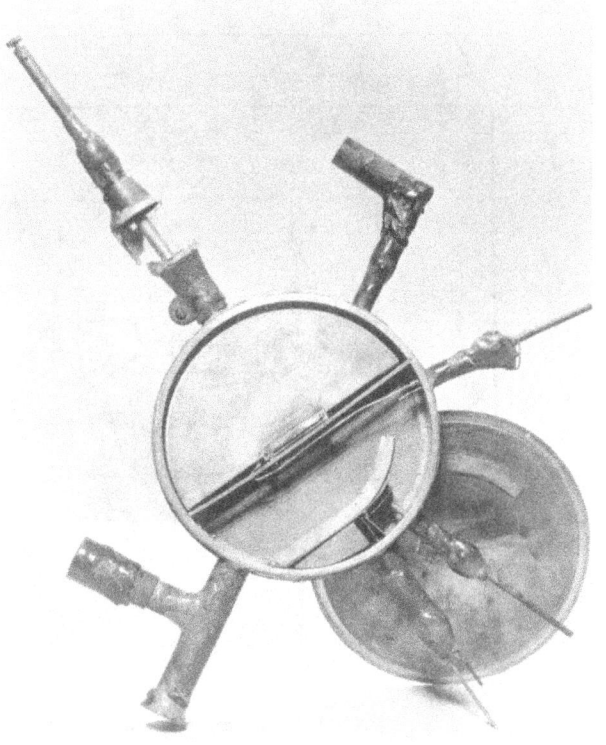

E.O.Lawrence,
M.S.Livingston
Berkeley, California

4 inch diameter
1 kV on the Dee
80 keV Protons

Ringcyclotron (1974)

590 MeV Protons

8 Magnet à 250 Tons

4 Cavities à 500 kV

The original Ringcyclotron was proposed in 1962 by Hans Willax and J.P.Blaser

original Cyclotrons

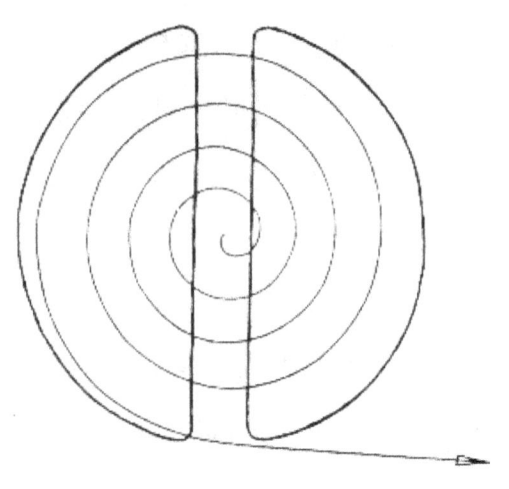

the cyclotron as seen
by the inventor

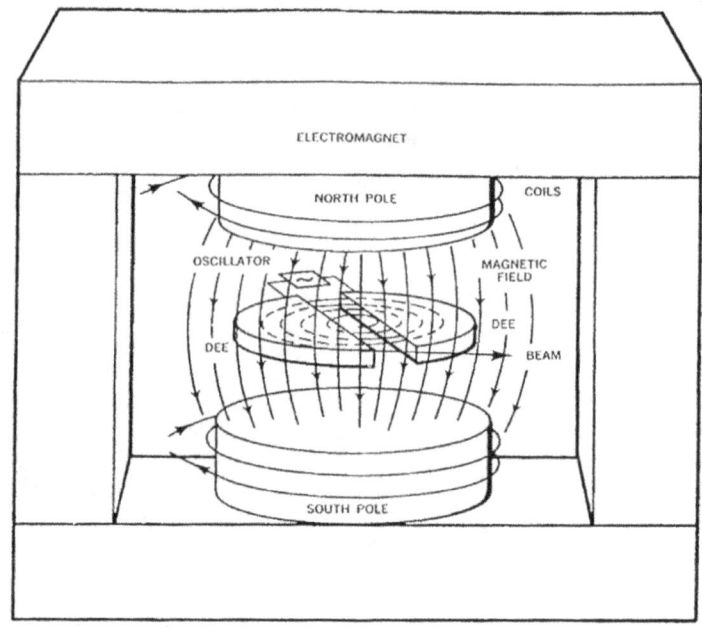

the first classical cyclotrons

Injector I Cyclotron

Philipps (1973-2011)
72 MeV Protons:
100-200 µA
polarized: 10 µA

Ions, Energy/Nucleon:
$E/A = (Z/A)^2$ 120 MeV
e.g. Deuterons, α :
30 MeV/Nucleon

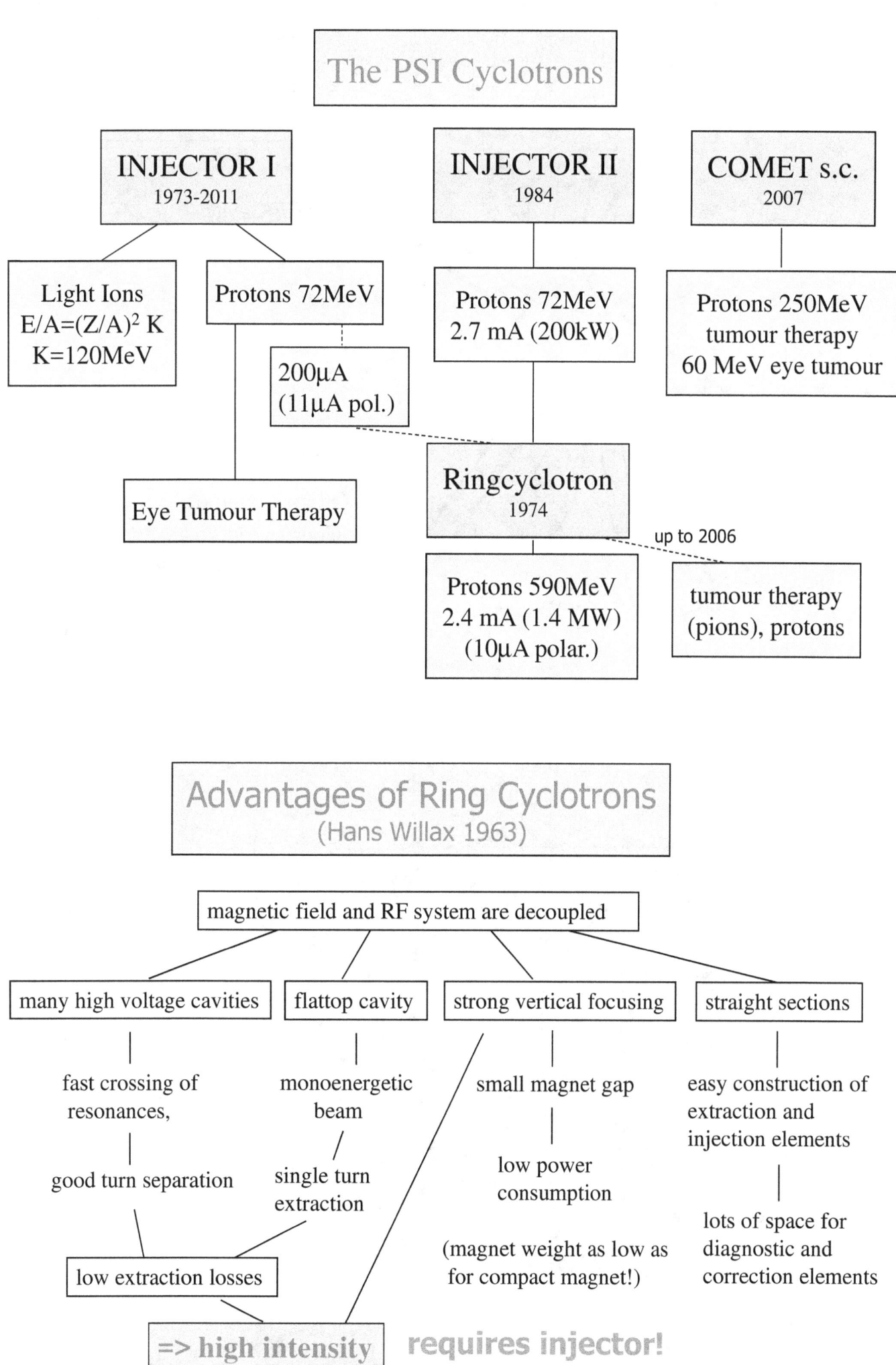

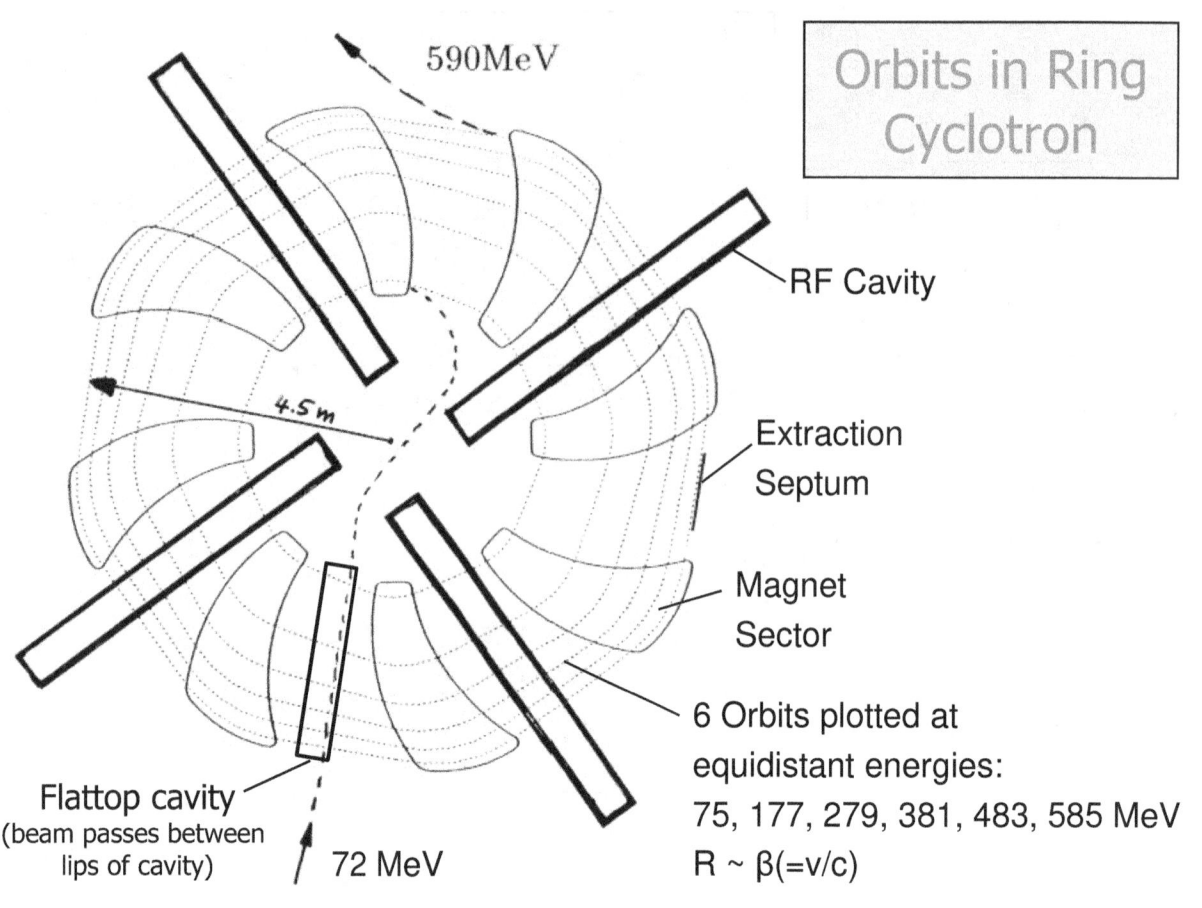

Injection of 870 keV Protons into Injector II

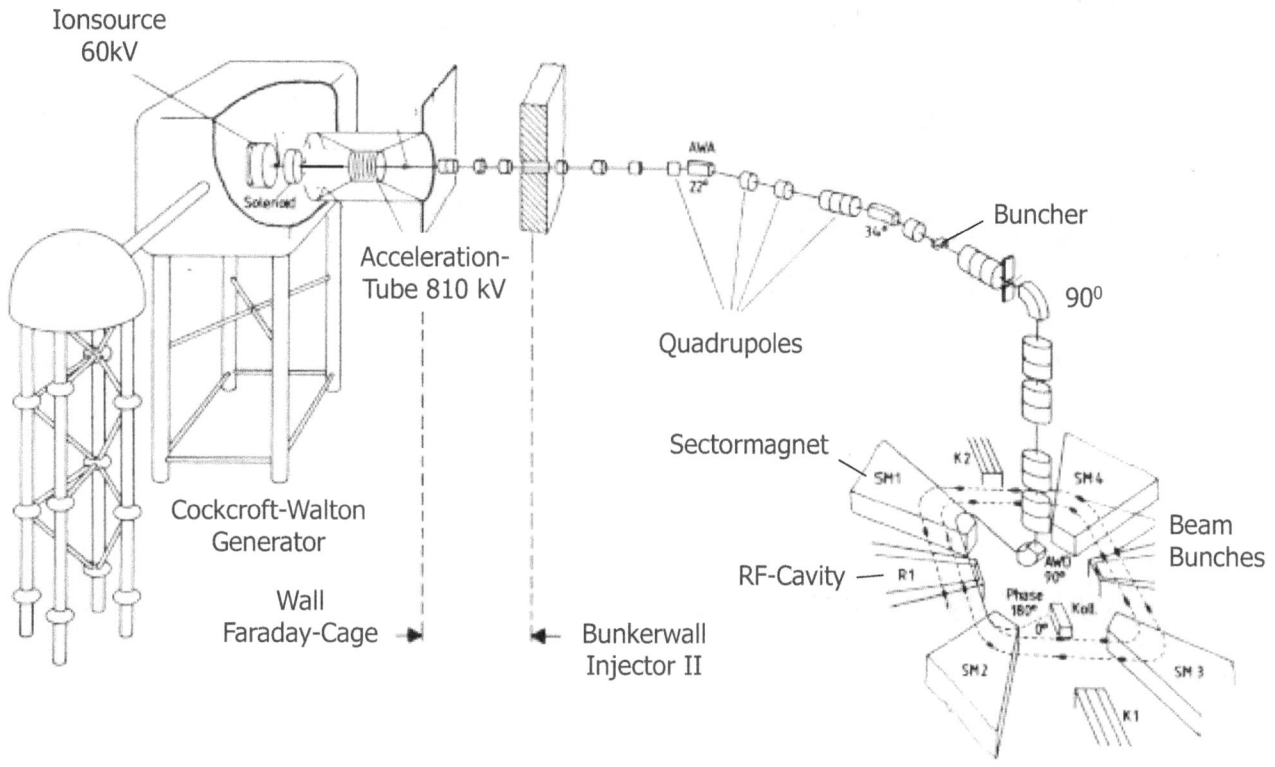

Injector II

Recipe for high Intensity

- continuos beam (cw)

- very low extraction losses

=> separated turns with
   large turn separation dR
   at extraction

=> high energy gain per turn,
   powerful RF-system with
   high voltage cavities

   dR ~ Radius R
=> **large machine radius !!**

the last 5 turns in the Injector II

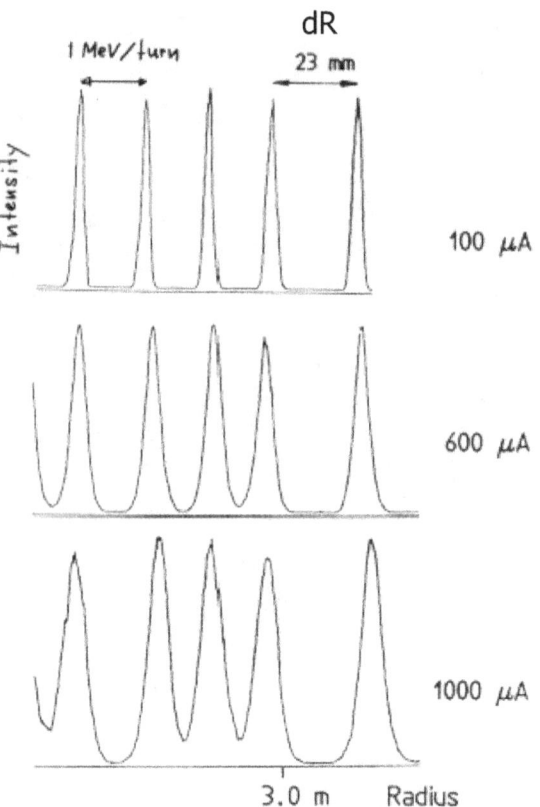

4 new Cu Cavities in Ringcyclotron (2008)

590 MeV Protons

1.4 MW Beam Power
(world record!)

4 Cavities à 850 kV

Extraction ≈ 99.98 %

New Copper Cavity (5.6m long)

50 MHz, CW

Voltage limit 1 MV

(old cavity 0.72 MV)

at 850 kV and 2.4 mA:

250 kW loss in cavity

350 kW goes to the Proton beam

Beam limit 3 mA ?

Flattop Voltage gives minimum energy spread

Flattop RF-Voltage with addition of a 3.harmonic

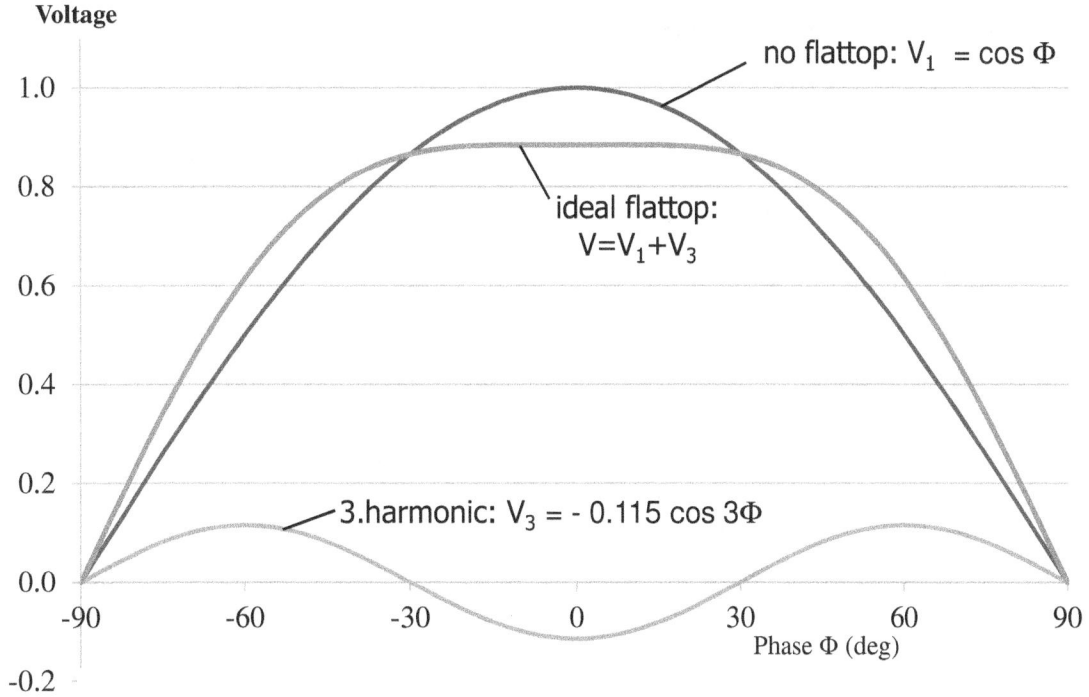

no flattop: $V_1 = \cos\Phi$

ideal flattop: $V = V_1 + V_3$

3.harmonic: $V_3 = -0.115 \cos 3\Phi$

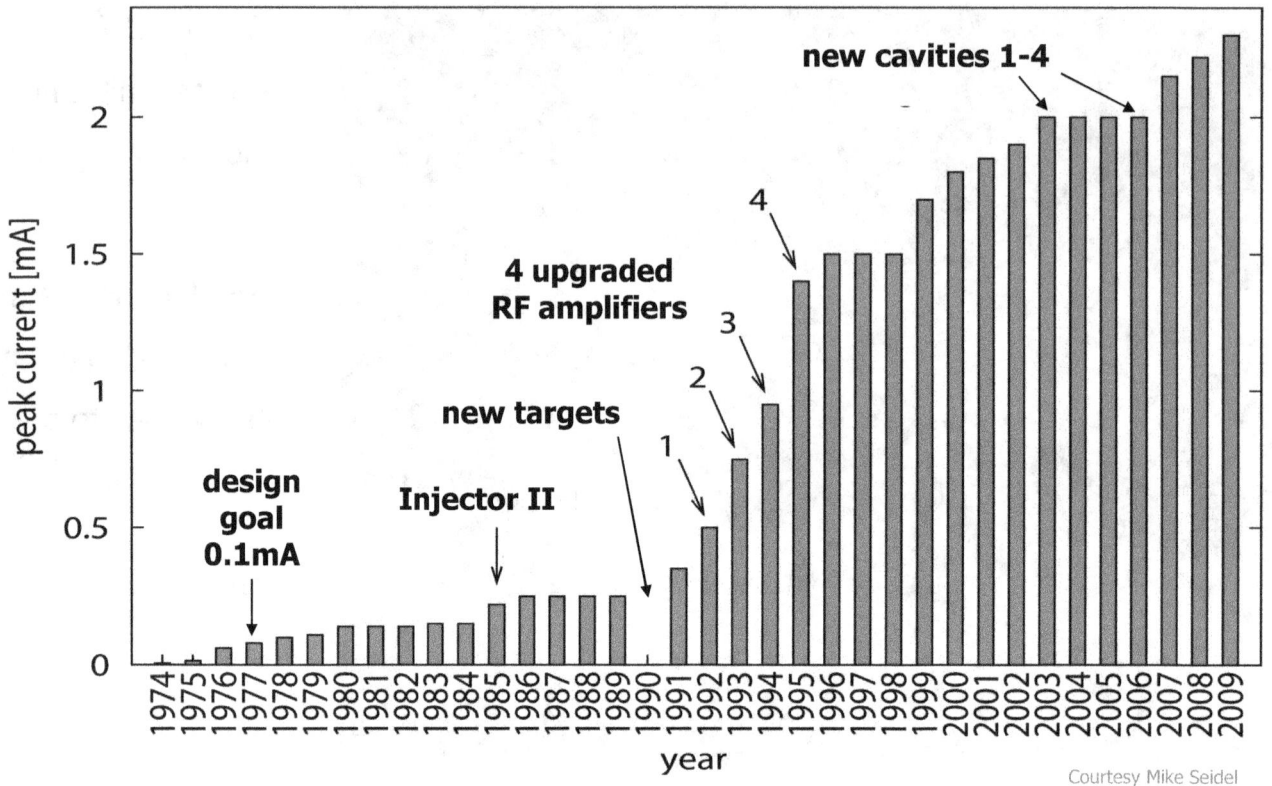

Current Limits in the 590 MeV Ring Cyclotron

- Goal of 100 µA for a **meson factory** was considered "very ambitious" by outside experts; at that time the record from synchro-cyclotrons was 0.1 µA !

- The concept of the Ring Cyclotron with separate sectors by **Hans Willax** created the potential for „high currents" with low losses due to high voltage cavities, large radius and strong vertical focusing

- critical point was Injector; final limit for original Injector I from Philips was 200 µA => Concept for Injector II was created before ring cyclotron was commissioned !

- In 2013 the record intensity was pushed to 2.4 mA, a **factor of 24 over its design goal !**

Key points for PSI Injector Cyclotrons

Philipps Injector (1974-1994)

- **vertical collimators** in center of cyclotron
  give excellent beam quality (Th.Stammbach)

  => extraction efficiency jumped from 70% to 94% (100µA => **200µA**)

Injector II (since 1995)

- space for new Injector II was **foreseen in initial layout**
  large size of Injector II allows injection from **Cockcroft-Walton** at 870 keV
- large size gives **large turn separation** at extraction => very low losses
- „**spaghetti effect**" increases space charge limit by a factor of about 10 to ≈ **3 mA**

Current Limit

- **RF-System**: today 1.4 MW are delivered to the beam
  (100 kW are taken out again by the flattop system).

  => any further increase in current requires another upgrade of the RF system.

- **Activation of Cyclotron components**:

  Initially a loss of 5 µA at extraction was considered as acceptable !?
  Today the tolerance is about 0.5 µA in order to allow "hands-on" maintenance.

  => The losses at extraction determine the current limit

- **Transversal space charge forces**

  defocusses the beam: The vertical tune is lowered and the beam size increases.
  For the Ring Cyclotron this effect becomes serious for currents above 10 mA

- **Longitudinal space charge forces**

  much more serious, because they increase the energy spread and thus
  the final beam size at extraction, increasing beam losses.

Space Charge Fields in Sector Model

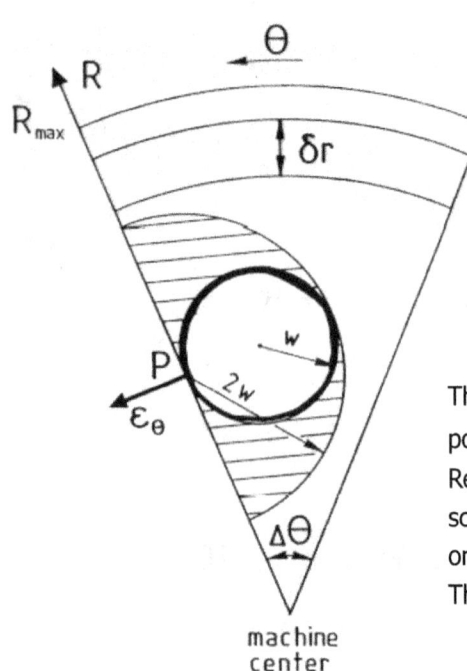

top view / machine center

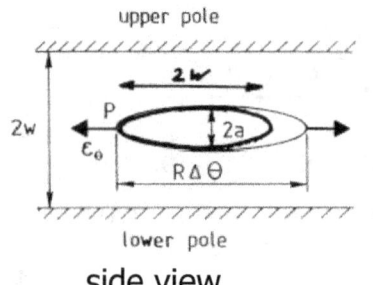

side view

circulating protons fill a cake-like piece with azimuthal extension $\Delta\theta$. Neighbouring orbits are assumed to overlap radially.

The azimuthal electric field at the edge of the „piece of cake" at point P is approximated by the calculable field of a **Disc** with radius w. Reasoning: the charge of the protons outside of the half circle around P is screened by the upper and lower poles and protons in the hashed area give only a small contribution to the azimuthal field ϵ_θ.

The proton at P gains through ϵ_θ an additional energy/turn:

$$dE/dn = 2\pi R\, \epsilon_\theta$$

This simple model predicts, that the intensity limit from longitudinal space charge forces increases with V^3 !!
(V=cavity voltage/turn)

Longitudinal Space Charge

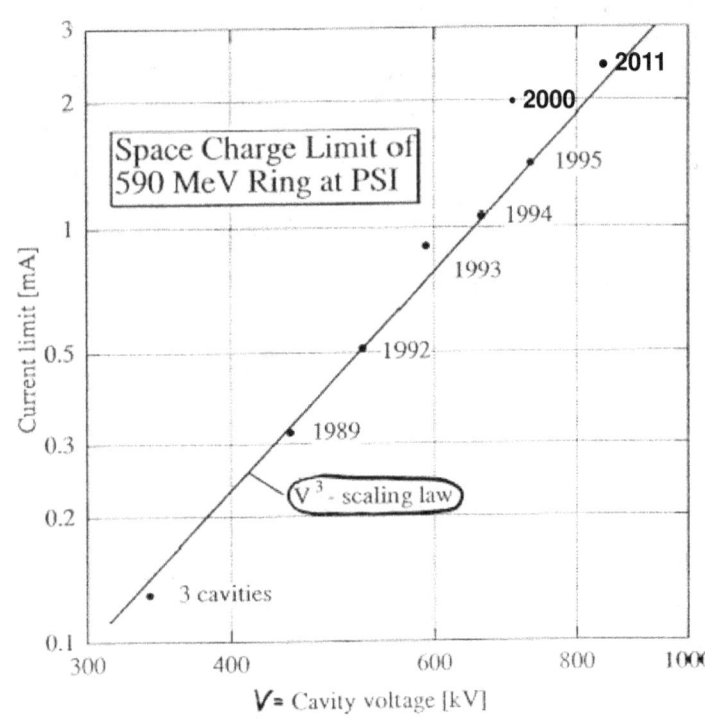

Longitudinal space charge **forces** increase the energy spread
=> higher extraction losses
=> limit on beam current

Remedy:
higher voltage V on the RF cavities
=> lower turn number **n** (V·n = const.)

current limit ~ V^3 ~ $1/n^3$!

There are 3 effects,
each giving a factor V(~1/n):

1) beam charge density ~ **n**
2) total path length in the cyclotron ~ **n**
3) turn separation ~ **V**

W.Joho, 9th Int. Cyclotron conference
CAEN (1981) p.337

Longitudinal Space Charge in a Cyclotron Bunch

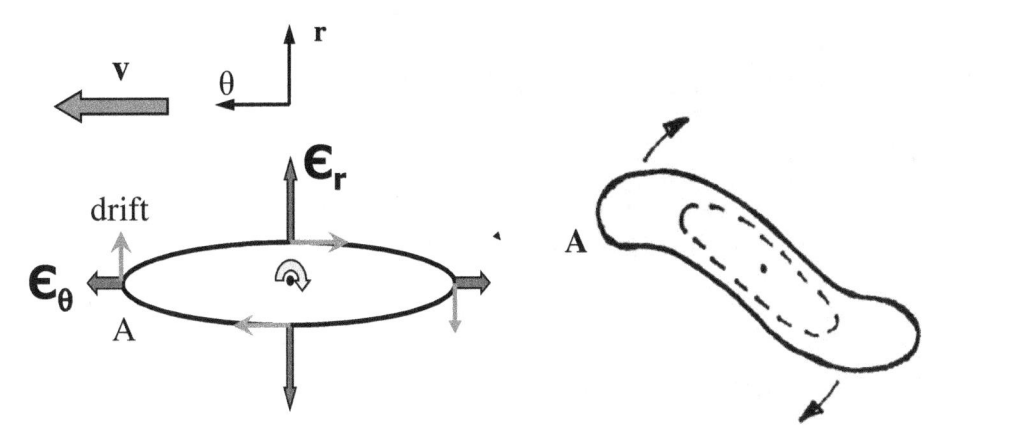

Particle at position A:

=> gains additional energy from space charge forces

=> moves to higher radius due to isochronous condition

=> rotation of the bunch

=> nonlinearities produce spiral shaped halos

=> production of a **rotating sphere** (mixes phases)

Longitudinal Space Charge in Injector II Cyclotron

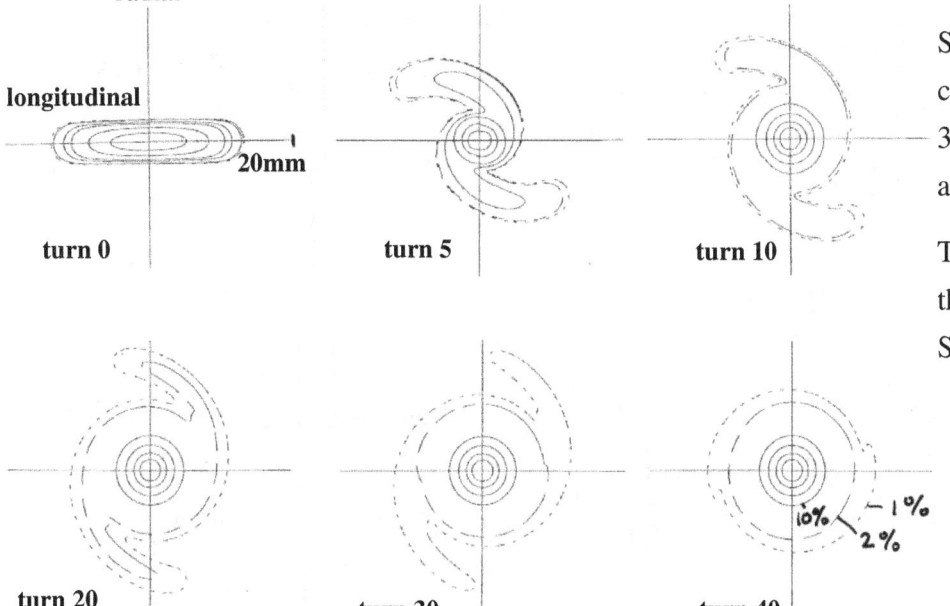

Simulation of a 1mA beam, circulating in Injector II at 3 MeV for 40 turns without acceleration.

The core stabilizes faster than the halos (calculations by Stefan Adam)

Model of the last turns in the 590 MeV Ring Cyclotron

- the turn separation is proportional to the orbit radius R and the cavity voltage V

  => concept of a large ring cyclotron
  with many high voltage cavities

- Flattop cavity gives mono-energetic beam

  => leads to single turn extraction

  => an eccentrically injected beam is still eccentric at extraction.

  => This can be used to increase the radial separation between the last two turns.

- In our simple model we assume an average turn separation of 6mm at extraction (energy gain 3 MeV/turn)

The Miracle of the ideal horizontal Tune Qr in the Extraction Region

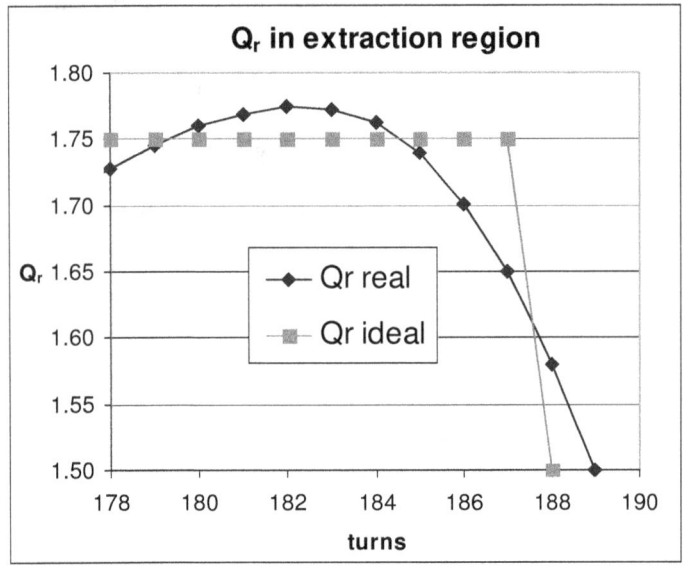

In the **ideal case** one can, with a tune of 1.75, overlap 3 turns, while with a fast drop to 1.5 the last turn is pushed away from the previous 3 turns.

In the fringe field region of the PSI ring cyclotron the **real tune** is close to the ideal one (just by pure luck!), giving an increase of the last turn separation from 6 to 17mm, using eccentric injection. The drop of Qr in the fringe field helps to increase the turn separation even for a centered beam (dR ~ $1/Q_r^2$)

The fringe field of the last sector magnet before extraction defocuses the beam horizontally. To compensate this a so called "Panofsky quad" in front of this last sector magnet is vital (proposed by George Vecsay).

the last turns in the Ring Cyclotron, model with real tune Q_r

590 MeV Ring Cyclotron last 9 of 188 turns for a 2 mA beam

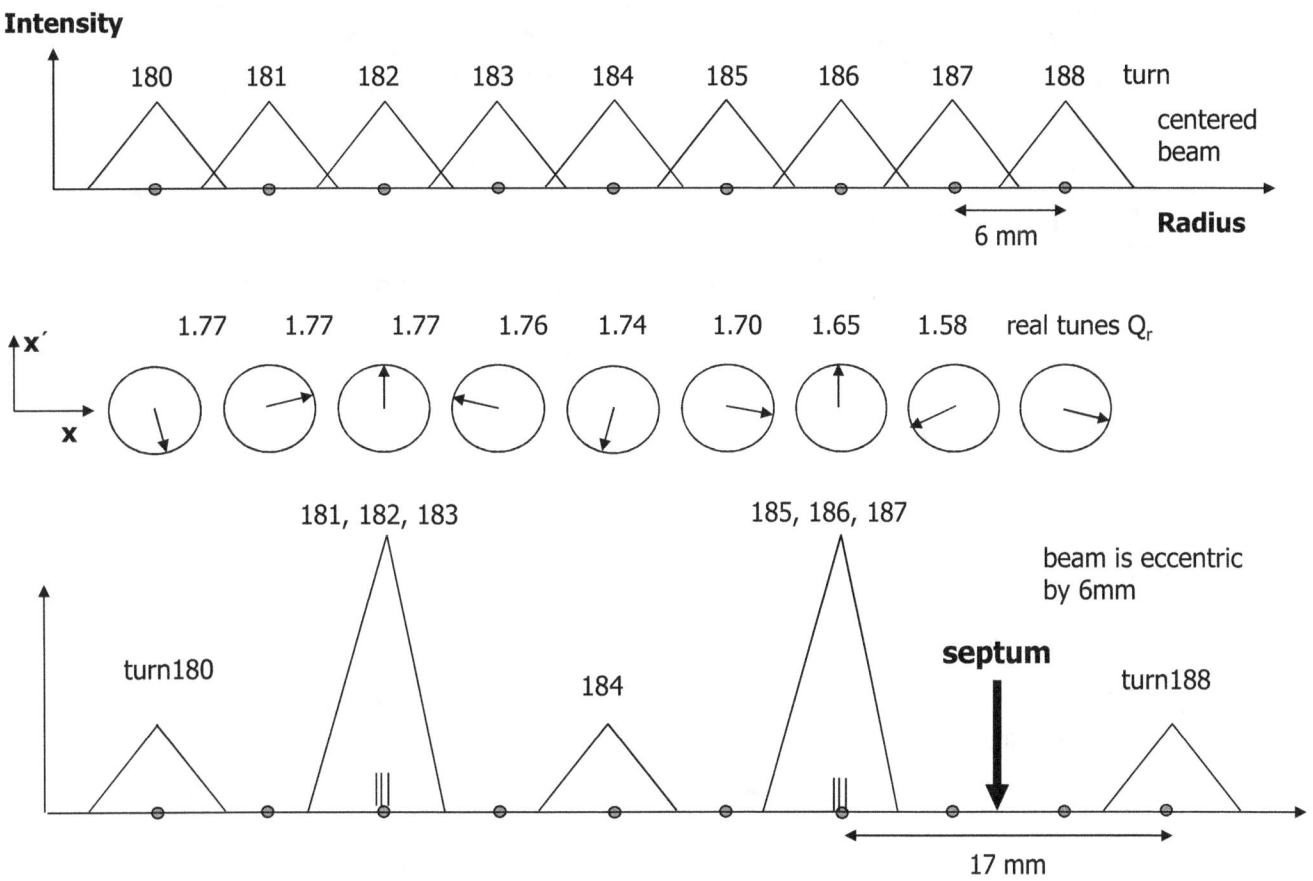

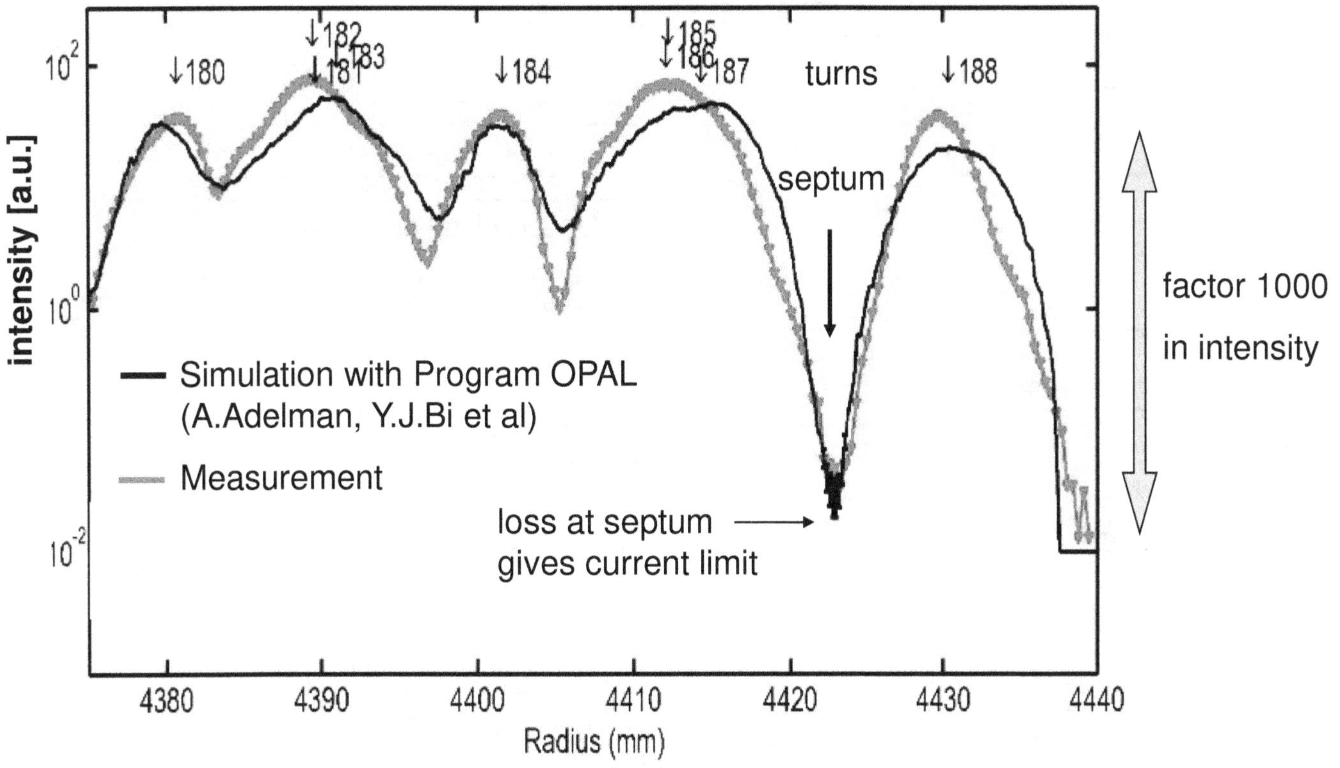

Reduction of Extraction losses

Cavity Voltage [kV]	Beam	Flattop Cavity	Losses [µA]	Losses [%]	Imax [µA]
500	Orig. Design	no	5 *	5	100
450	centered	no	0.5	1.2	40
450	eccentric	no	0.5	0.25	200
450	eccentric	yes	0.5	0.1	500
850	eccentric	no	0.5	0.06	800
850	eccentric	yes	0.5	0.02	2'400

Loss Reduction by	Factor	
Beam Quality	≈ 4	
eccentric Injection	≈ 5	
Flattop Cavity	≈ 3	
Cavity Voltage	≈ 5	($I_{limit} \sim V^3$)
total	300	

* Losses up to 5µA at 590 MeV were originally thought to be handable. The presently chosen limit of 0.5 µA allows manageable maintenance during shutdowns, avoiding the use of robots.

Success Factors for PSI Ring Cyclotron

1. Magnets and RF-System are decoupled
2. 4 high voltage cavities → $I\ max \sim V^3$
3. large Radius R → high turn separation $dR/dn \sim R \cdot V$
   → fast acceleration into fringe field, where Qr drops: $dR/dn \sim 1/Q_r^2$
4. excellent beam from Injector → separated turns
5. Flattop Cavity → high phase acceptance ΔΦ
6. eccentric Injection → Qr at extraction drops from 1.75 to 1.5
   → wins factor 3 in dR/dn: 6mm => 17mm
7. straight electrostatic Septum with 0.1mm strips → Losses at extraction ≈ 10^{-4}
8. Continuous Beam (CW) → 1.4 MW Beam Power

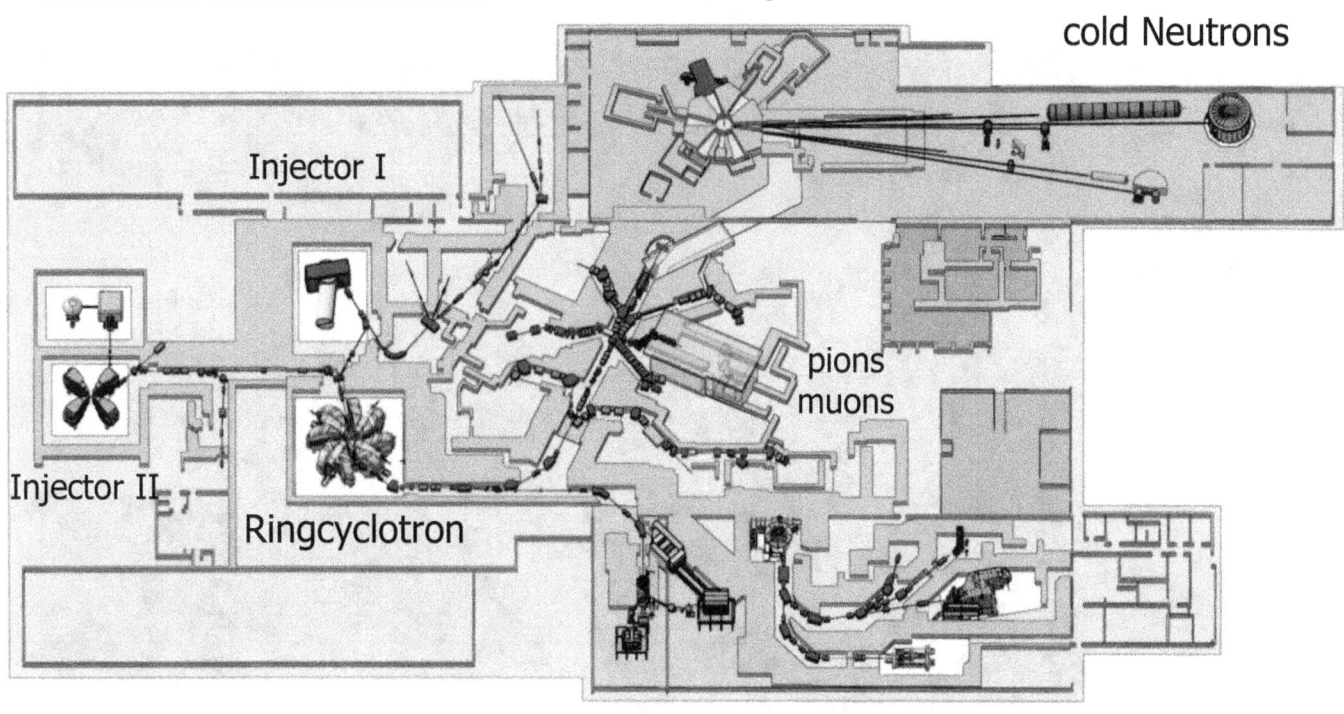

"slow" Neutrons for Material Research

- Production of fast Neutrons
- slowing down in Moderator

1. Fission of Uranium (U^{235}) in a Reactor

2. Spallation of heavy Nuclei (e.g. lead) by Bombardment with Protons from an Accelerator

   => safe and fast turning off !

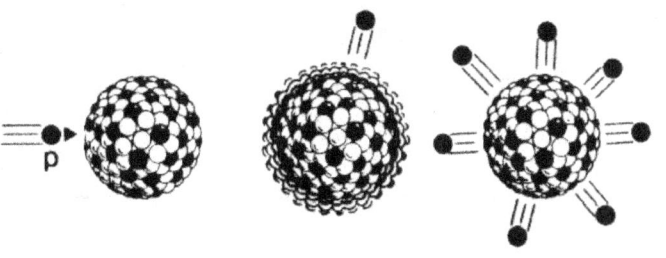

SINQ Neutron Spallation Source

cold Neutrons

Proton (590 MeV) => Lead Nucleus => ca. 10 Neutrons
=> Moderation to < 0.025 eV => Diffraction on Material Probes

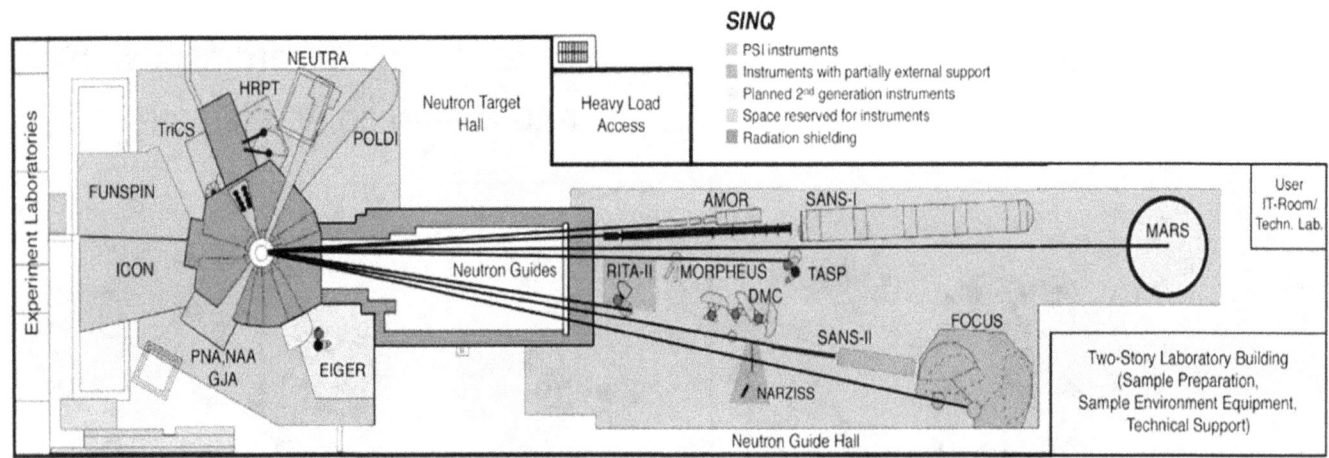

Neutron Scattering

Neutrons are scattered by a crystal
The arrangement of the atoms
can be reconstructed.

The wavelength has to match
the size of the structure:

Distance between atoms: ca. 0.1 nm
Wavelength of neutrons: ca. 0.1 nm

- Dysprosium
- Nickel
- Boron
- Carbon

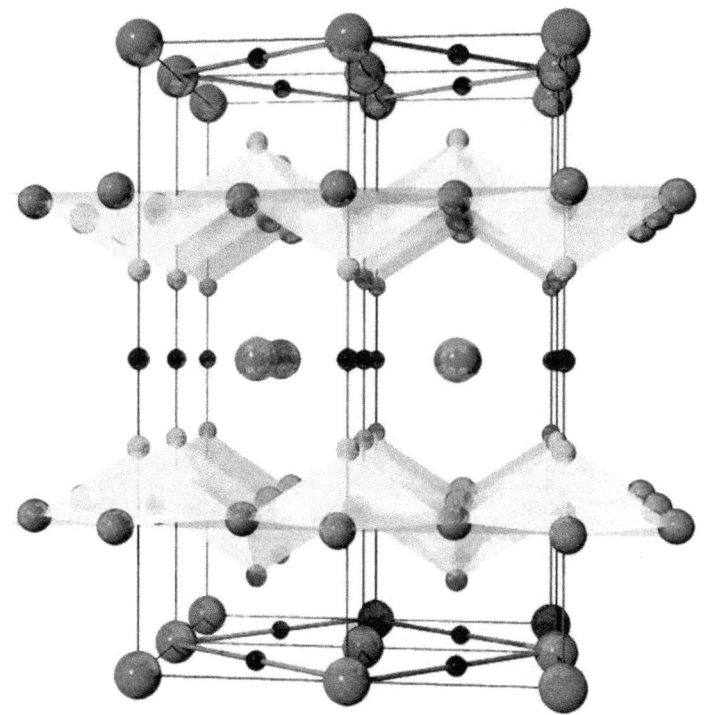

Radiography with Neutrons

=> the interior of big objects becomes visible

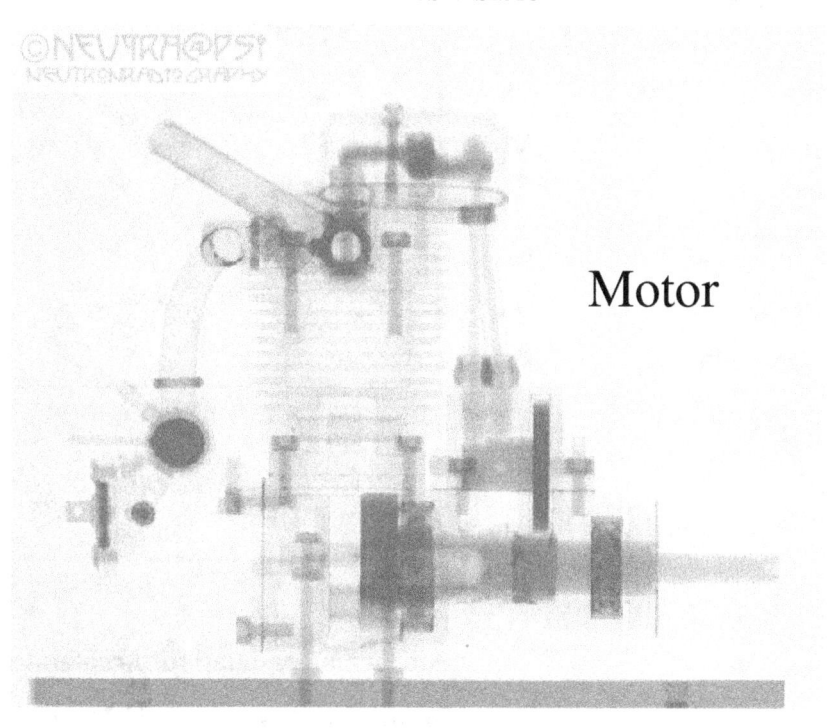

Motor

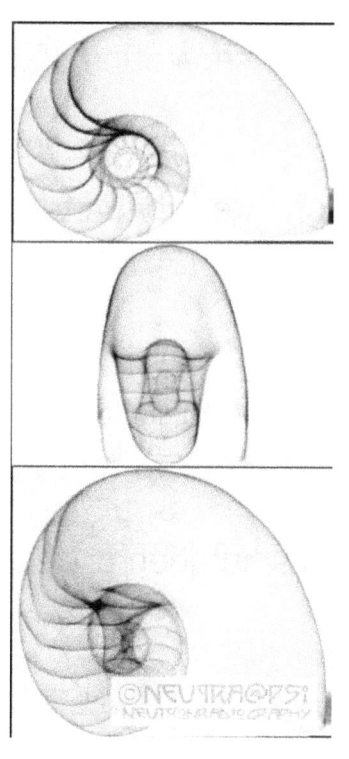

Energy Amplifier Concept (C.Rubbia)

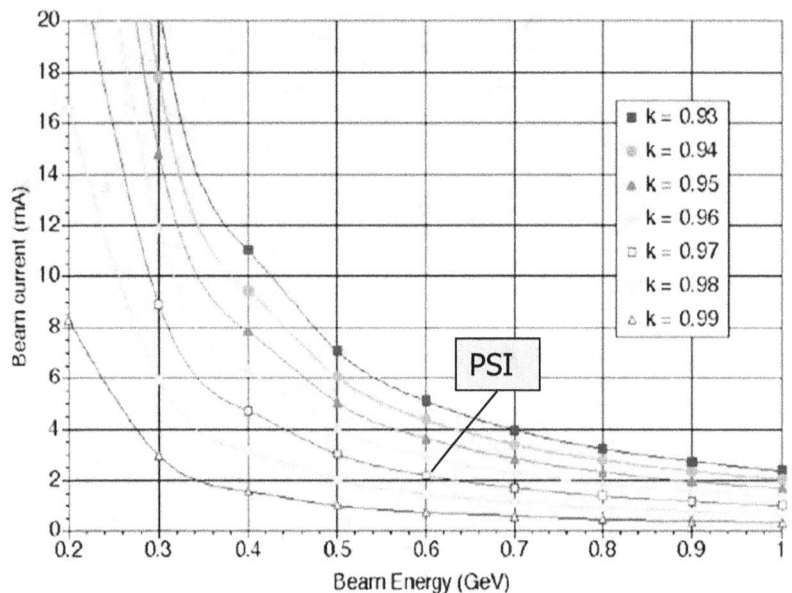

Beam current needed to produce 80 MW$_{th}$ with protons for different criticality factors k (ref. Ansaldo 2001)

example: 600 MeV PSI Cyclotron
in future with 3 mA => **1.8 MW**
=> production of neutrons in
   subcritical reactor (e.g. k=0.97)
=> 110 MW$_{th}$ => 40 MW$_{el}$

=> power plant with 1 GW$_{el}$ needs
35 mA protons at 1 GeV => s.c. Linac

- inherently safe
- use of Thorium (big reserves)
- no production of Plutonium for weapons

reduction of lifetime of nuclear waste !

chem. separation of long lived actinides

⇩

high intensity proton beam
≈ 40 MW (30*PSI), ca. 2050 ?

⇩

production of neutrons

⇩

transmutation of actinides

Reduction from 1 Mill. years
to 400 years

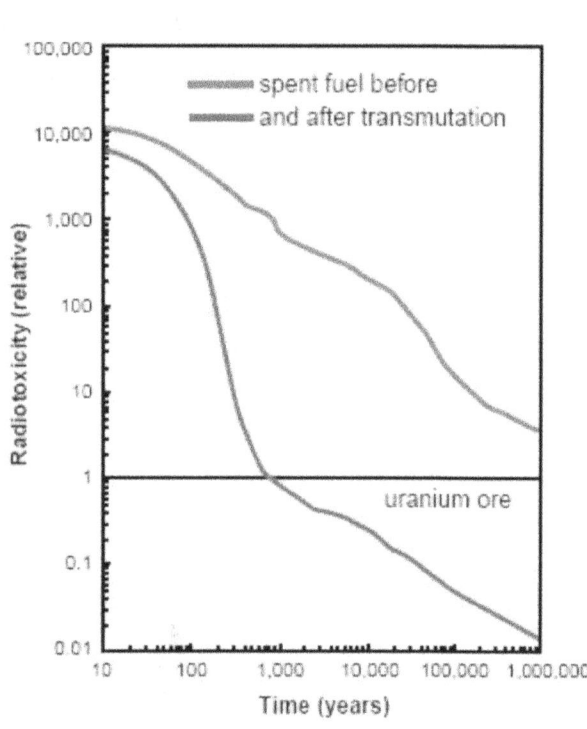

European Roadmap for Accelerator Driven Systems ... ENEA Italy 2001

History of the Cyclotron

1929	Idea by E.O.Lawrence in Berkeley (inspired by R.Wideroe!)		
1931	4 inch cyclotron	80 keV	p
1932	10 inch cyclotron	1.2 MeV	p
1934	26 inch cyclotron	7 MeV	p
1939	60 inch cyclotron	16 MeV	d
1946	184 inch synchrocyclotron	200 MeV	d
		400 MeV	α
1938	Idea for sectored cyclotron (AVF) by Thomas		
1962	88 inch sector cyclotron	K=160 MeV	ion
1974	SIN/PSI Ringcyclotron	590 MeV	p
1982	supercond.cyclotron MSU	K=500 MeV	ion

2008: ca. 90 indiv. cyclotrons, ca. 200 commercial cyclotrons

Differences between Cyclotrons and Synchrotrons

Cyclotron

- The orbits are spiraling outwards, the radius increases proportional to the velocity
- the revolution frequency is constant (isochronous)
- The magnetic field is constant in time
- The beam is continuos (cw, with RF frequency)

=> ideal for high intensities and high power

1.4 MW at PSI

Synchrotron

- the orbit remains constant at a fixed radius
- the revolution frequency changes
- the magnetic field increases in time
- the beam is pulsed (≤ 50 Hz)

=> ideal for high energies

7 TeV at CERN, Geneva

Analogy with transport of people:

rolling staircase (metro, shopping center)   elevator (skyscraper)

Ring Cyclotron September 1973

1 Hans Willax
2 Miguel Olivo
3 Thomas Stammbach
4 Werner Joho
5 Christa Markovits

First 600 MeV Protons on Target 25.2.1974

Richard Reimann
Manfred Daum

Thomas Stammbach   Werner Joho   Paul Rudolf   Hans Willax   Urs Schryber   Jean Pierre Blaser
Francesco Resmini

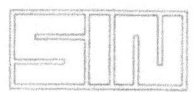 The Dream Team of 1974

Old Sinners Villigen, 24. 2. 1974

        Gerber   Blaser   Willax   Joho   Schryber   Lanz
Daum   Olivo   Steiner   Frosch   Tschalär

Courtesy Manfred Daum

Int. Cyclotron Conference 1975 Zürich

Jean Pierre Blaser Director — Hans Frei — Werner Joho — Stanley Livingston

SLS
(140m Diameter)

Swiss	**L**ight	**S**ource
Synchrotron	**L**ichtquelle	**S**chweiz
Source	**L**umière	**S**uisse
Sorgente	**L**uce	**S**vizzera

SLS Building

an architectural Juwel !   Team of Architects from Bern
(Gartenmann, Werren, Jöhri, Marchand)

Inside SLS

SLS interior Area

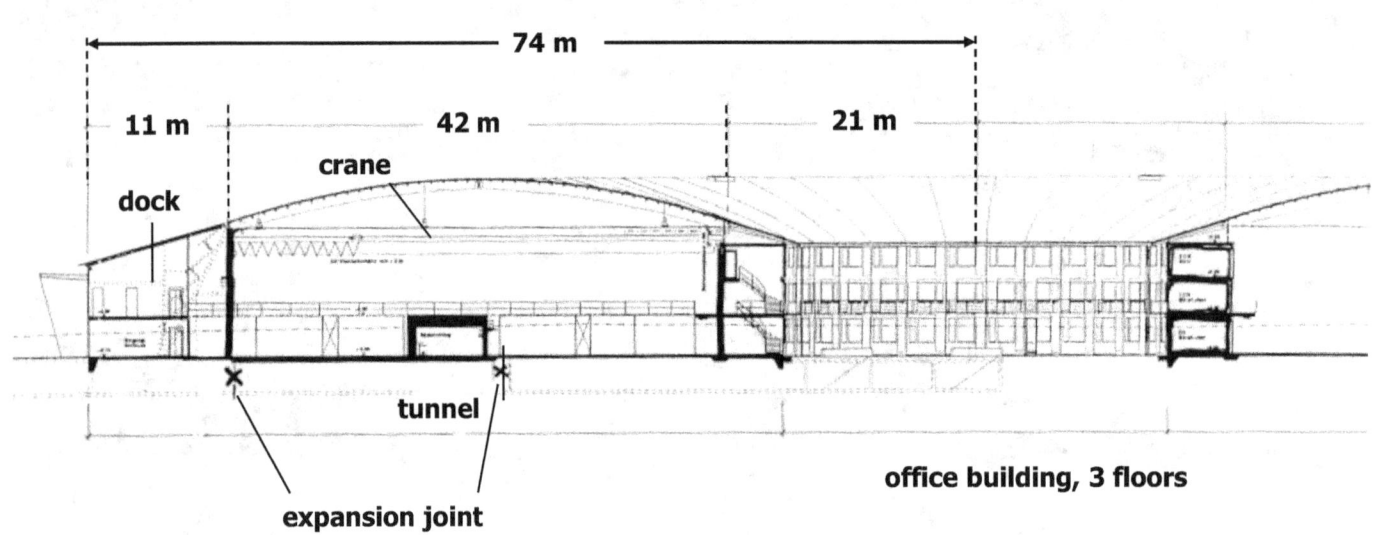

Building Concept

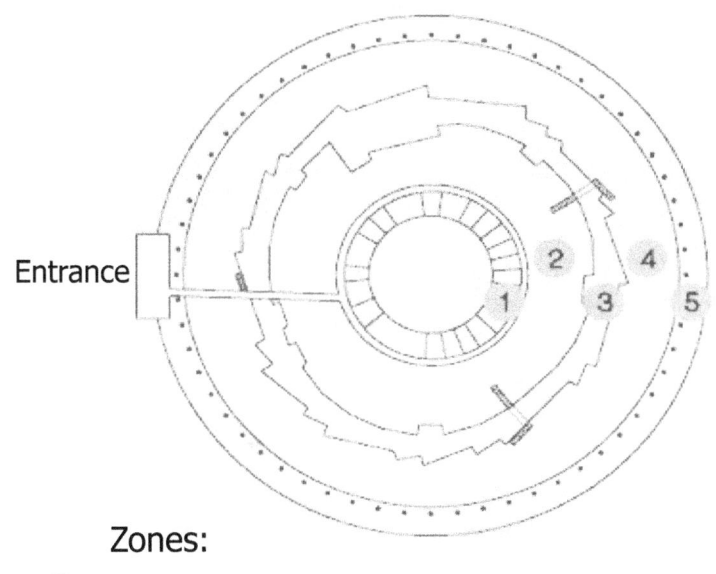

Zones:
1. Office Building (3 Floors)
2. Technical Galery
3. Tunnel (Storage Ring, Linac and Booster)
4. Area for Beam Lines
5. Outer Ring (60 Columns, Air Inlet System)

- separate annular Ring (40 cm) for Floor of Tunnel und Beam Lines (Zones 3, 4)

  => decouples Tunnel and Exp. Floor from rest of Building

- very stable Temperatures in Tunnel und Hall

  => **stable Conditions for Electron Beam and Beam Lines**

SLS Layout

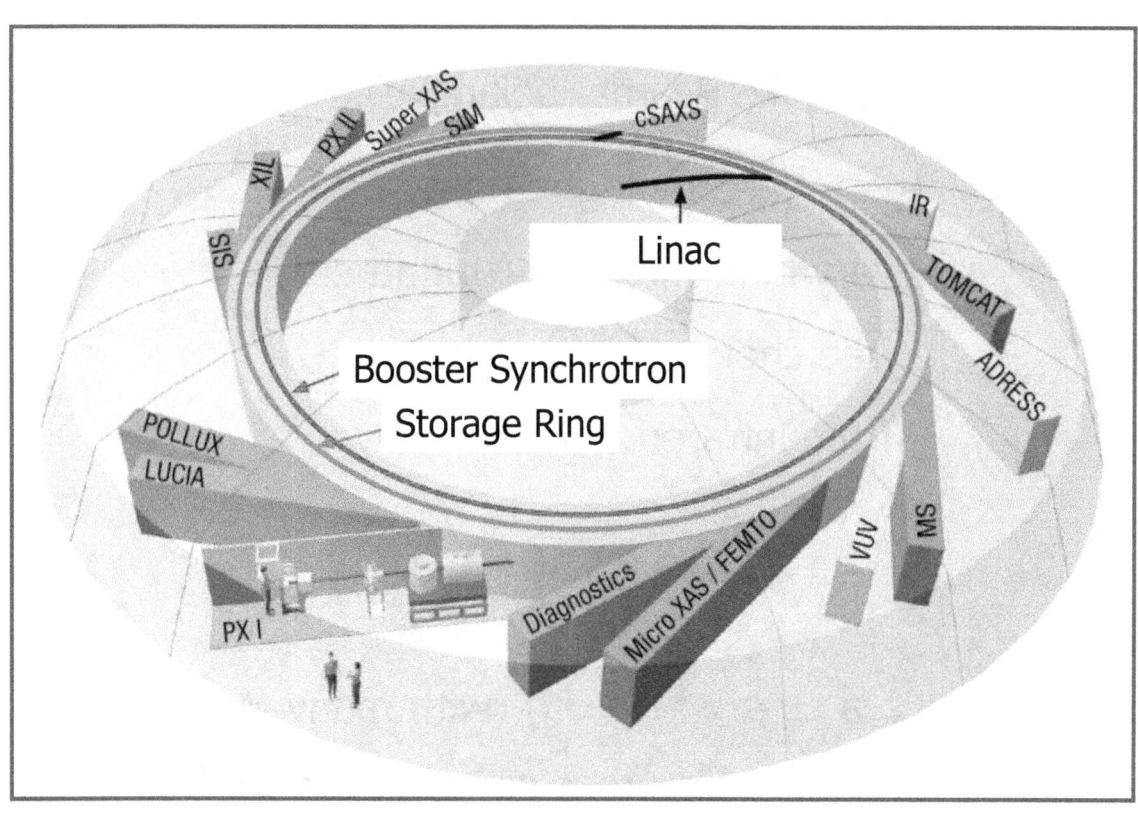

21 SLS Beamlines

SLS Beam Lines status Aug. 2015

| 02D TOMCAT S-bend 11/05 | 03M ADRESS UE44 04/07 | 03D PEARL Bend 06/13 | 04S MS > MS* W61 > U14 08/01 1/11 |

01D IRS Bend 08/07

12S CSAXS U19 05/07

04D VUV Bend 06/07

05L µXAS / FEMTO U19 / +W138 09/04 / 08/06

11M SIM 2xUE56 11/01

05D Diag / Optics Bend 07/01 / 05/07

10D SuperXAS S-Bend 05/07

06D PX-III S-Bend 10/07

06S PX-I U24 > U19 07/01 > 10/04

10S PX-II U19 01/05

9D Diag Bend 06/06

09L SIS/XIL 2xUE212 08/01

09LA SIS UE212V 424H 3/09

09LB XIL U70 6/09

8D Diag Bend 01/13

07D PolLux / nanoXAS Bend 11/06

07M PHOENIX / XTREME UE54 (ex-LUCIA) (10/04)10/09

sector
Name
Source:
planar Undulator
[in vacuum] [cryo]
elliptic Undulator
Wiggler
Superbend
normal Bend
start of operation

Courtesy A.Streun

Synchrotron Radiation

- laser-like Beams (polarised), generated by high Energy **Electrons**
- very high Intensity (Brightness)
- free choice of Wavelength from infrared to hard X-Rays

what is needed?

⇒ a Storage Ring (with many Magnets), where Electrons can circulate for hours

SLS Strategy

Quality
- high brightness, small emittance,
  ⇒ large circumference with many magnets

Flexibility
- large spectral range (VUV to hard x-rays)
- straights of 4 m, 7 m and 11 m => choice for undulators

Stability
- separation of building structure from floor
- stable temperature in tunnel and experimental hall
- positioning of the magnets on rigid girders
- fast orbit feedback (up to 100Hz) with high accuracy (< 0.2 µm)
- constant beam current with **top-up injection** (every 2.5 min)
  ⇒ constant heatload on optical components

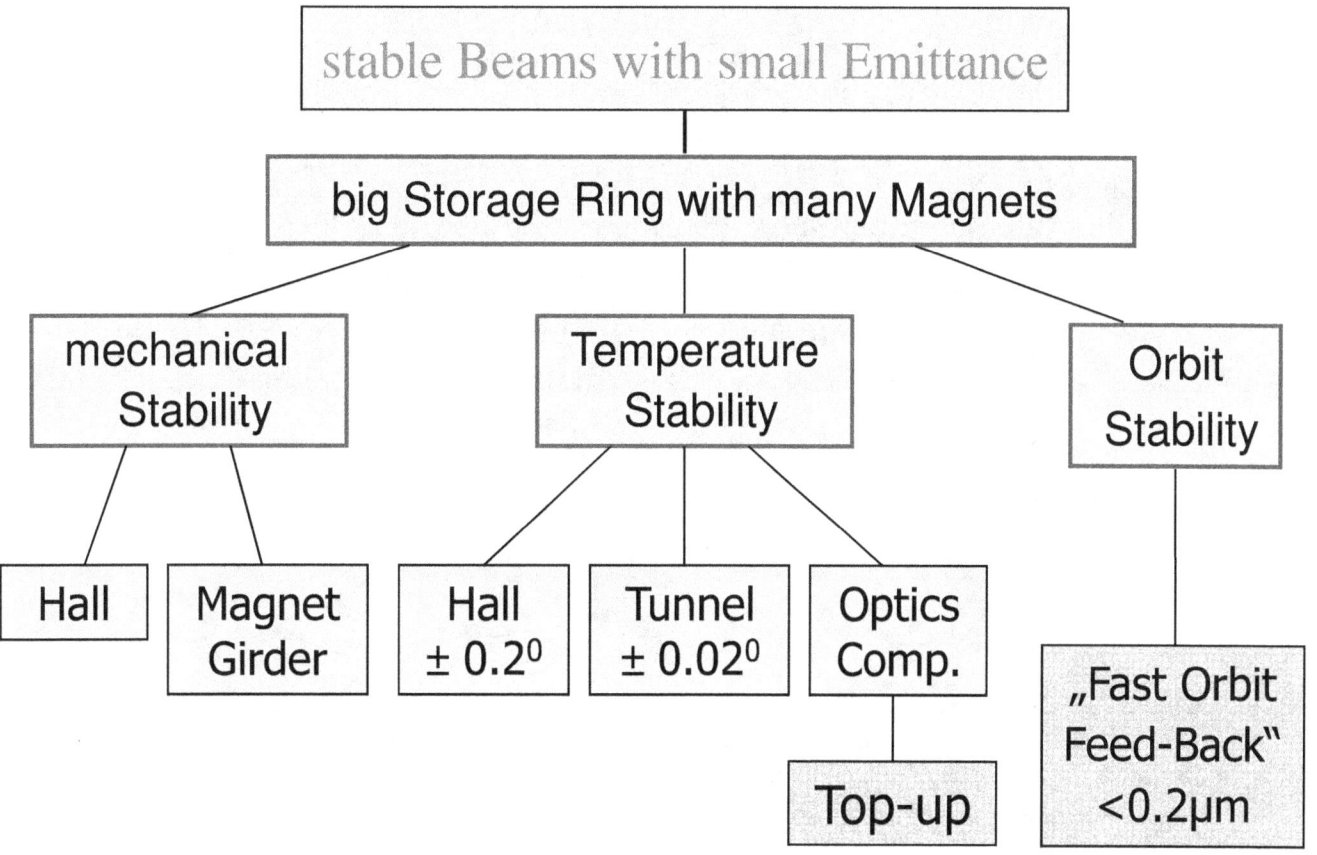

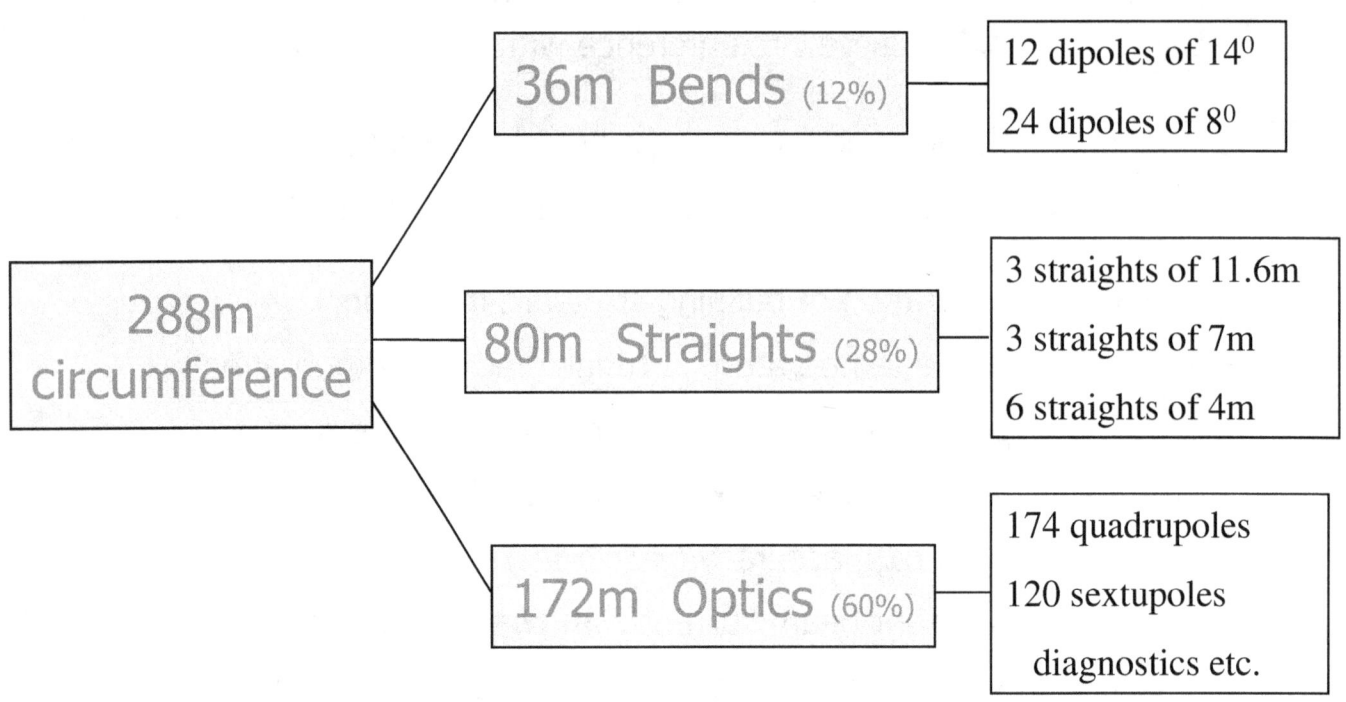

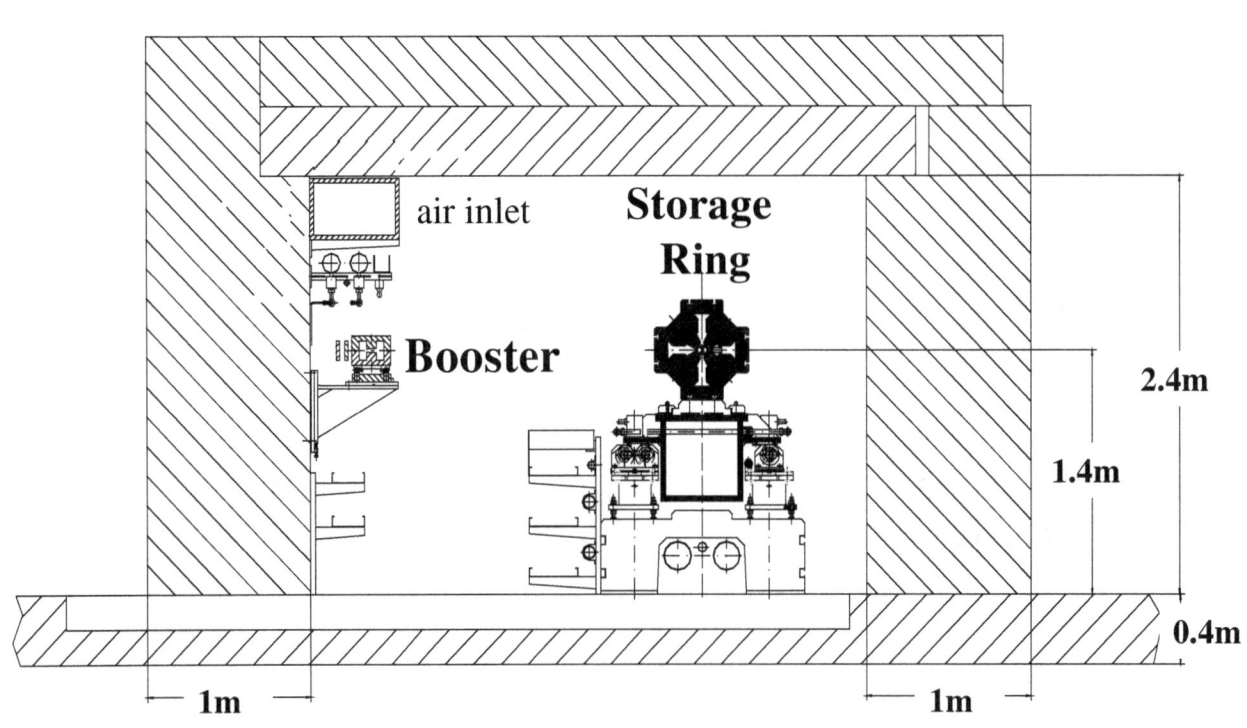

SLS Vacuum System

- electrons have to circulate for hours in storage ring
  => ultrahigh vacuum

- stainless steel vacuum chamber,
  preconditioned at 250^0 during 4 days

- constant cross section avoids beam instability

- Ante-chamber reduces desorption by synchrotron radiation

- vacuum chamber is connected to the girder only at the
  Beam **P**osition **M**onitors (=> floats inside the magnets)

SLS-Components

Accelerators

- Electron gun   90 keV
- LINAC   100 MeV
- Booster, 3 Hz   0.1-2.4 GeV
- Storage Ring, 288m   2.4 GeV

Beamlines

- Protein Cristallography
- Material Sciences
- Surface Microscopy
- Surface Spectroscopy
- environment sciences

Layout 100 MeV Linac

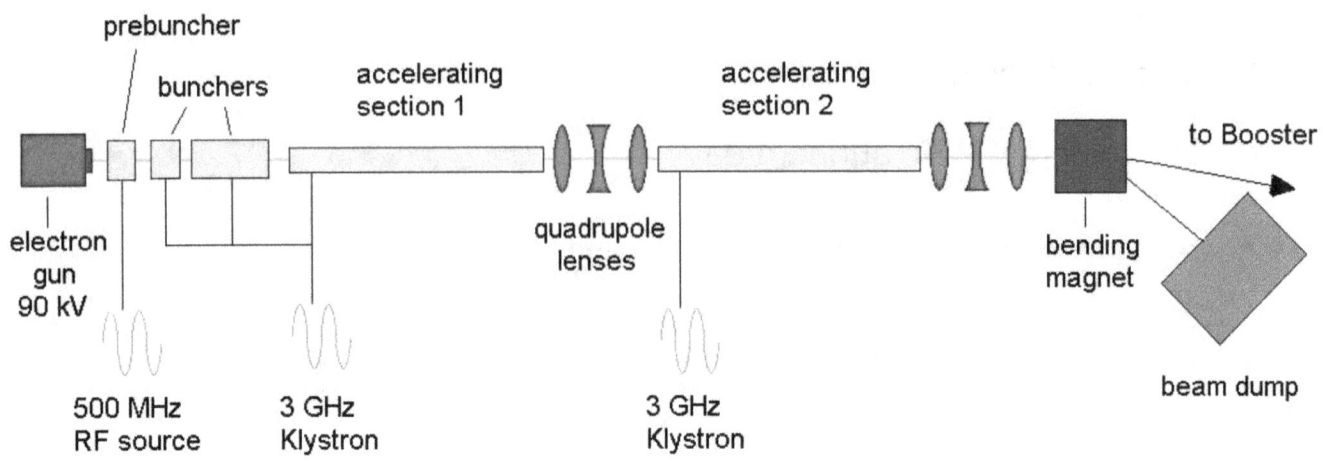

Linac

Electron Gun 90 keV

Linac 100 MeV

(11m long)

Injection Region

SLS Tunnel

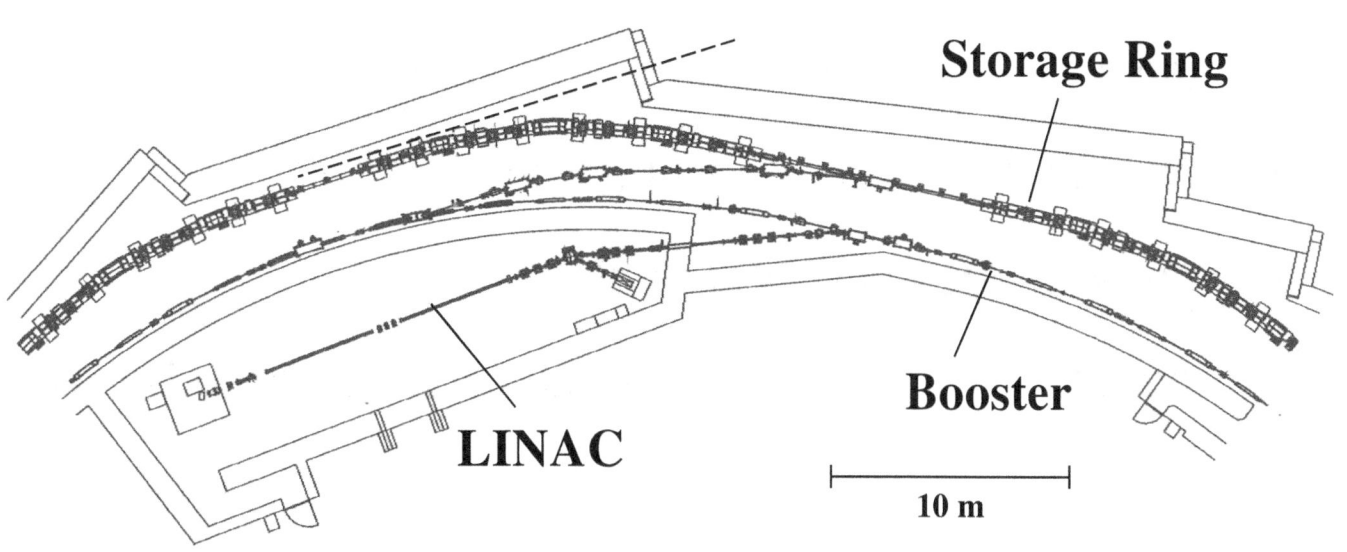

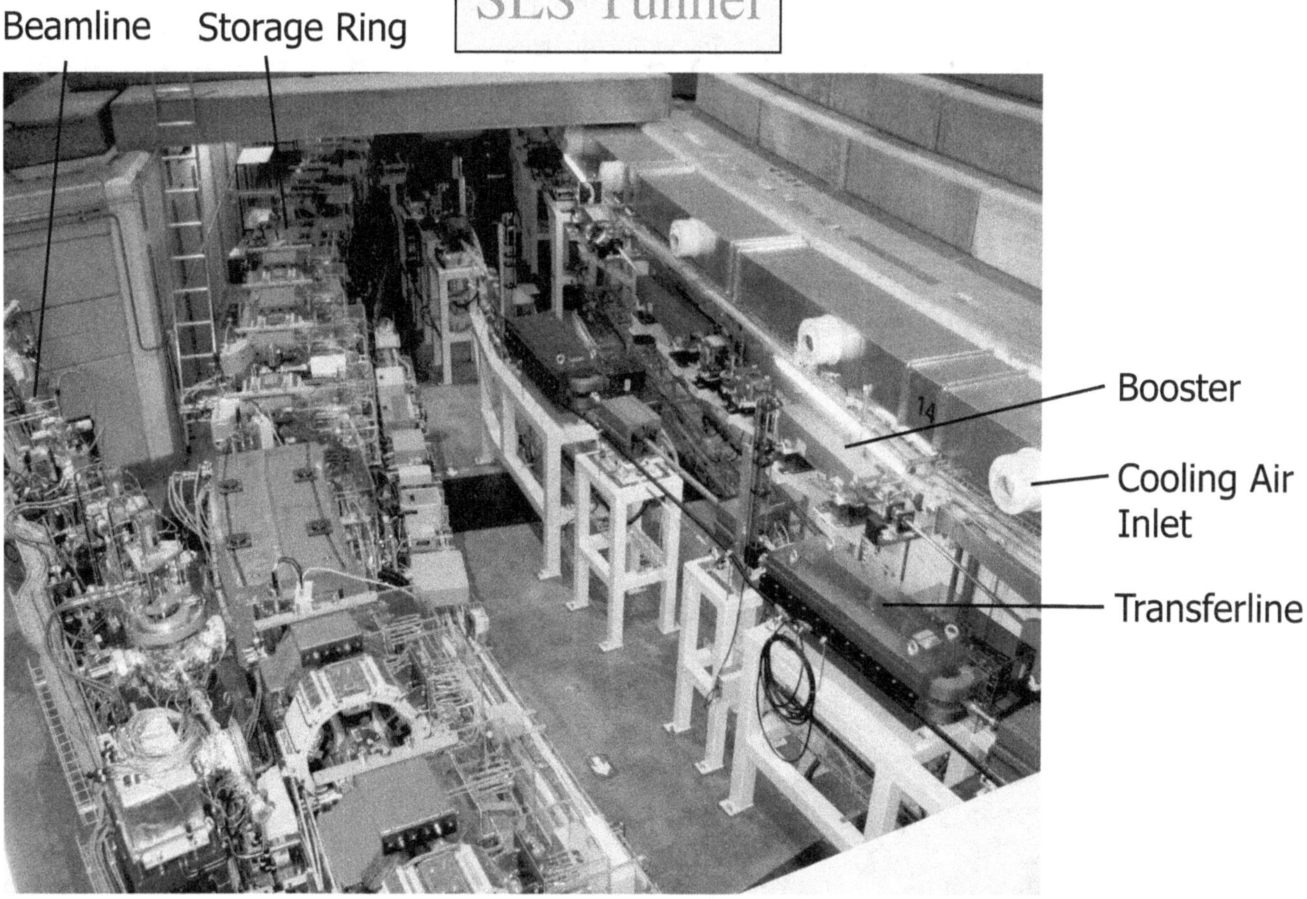

what is special at SLS ?

1. **very good beam quality, high brightness**
   (requires large circumference, many magnets)

   - microscopy (ca. 30 nm) , spectroscopy
   - small probes (microcrystals, 5-10 µm)
   - beamstability < 0.2 µm (with fast orbit feed back)

2. **large spectral range :**

   - infrared to hard X-rays (6 decades !)

3. **Short, Medium and Long Straights**
   (6 à 4 m, 3 à 7 m, 3 à 11.5 m)

   - flexibel undulator schemes

4. **Top-up Injection :**

   - constant beam intensity over days
     (beam lifetime is irrelevant !)

     => constant heat load on optical components

Booster Specialty

- **Booster in same Tunnel as Storage Ring** (G.Mülhaupt)

- => large Circumference
  => small Emittance

- efficient Injection into Storage Ring, filling in 6 min.

- compact, economic Magnets

- simple Vacuum Chamber (30 x 20 mm)

- **Top-up Injection**

- short refill (2s) every 2.5 min.
  => constant Beam Current

- => stable Temperatures on optical Components

- Energy Consumption < 20 kW !

Booster FODO-cell 5.4m long

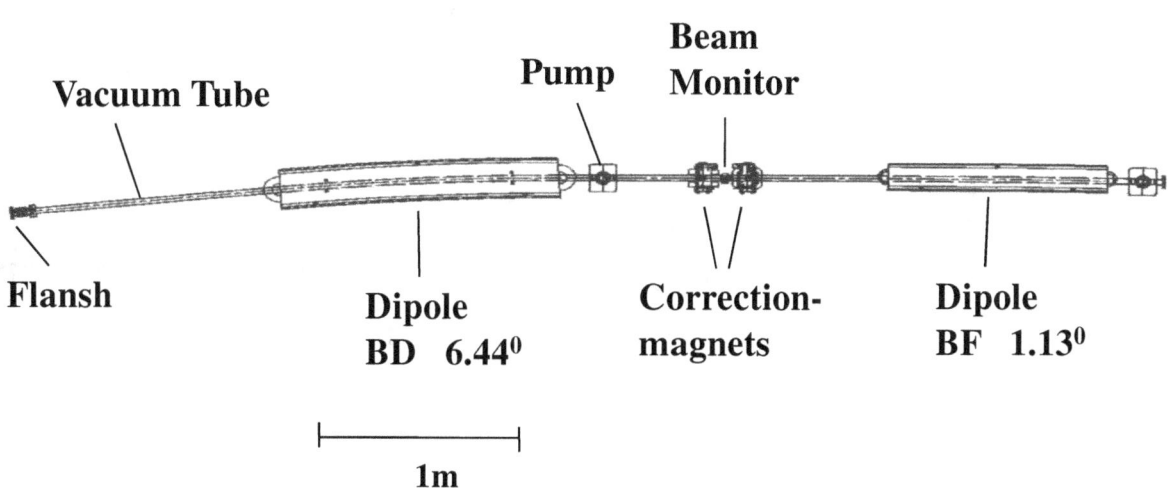

Vacuum Tube, Pump, Beam Monitor, Flansh, Dipole BD 6.44°, Correction-magnets, Dipole BF 1.13°, 1m

Booster

Acceleration in the Booster

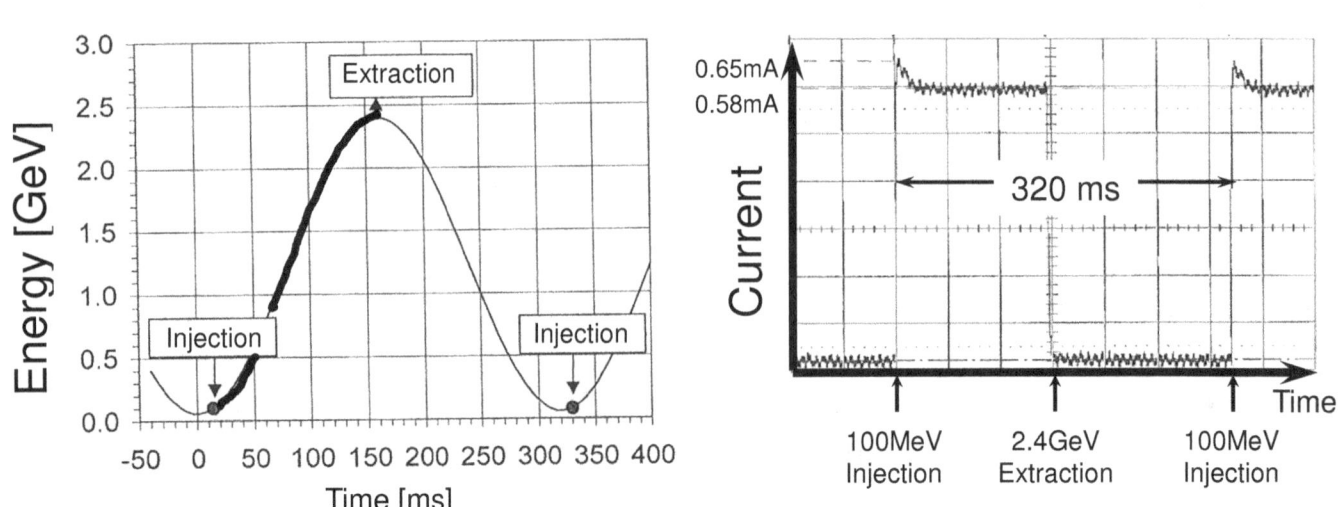

Lifetime vs. Top-up Injection

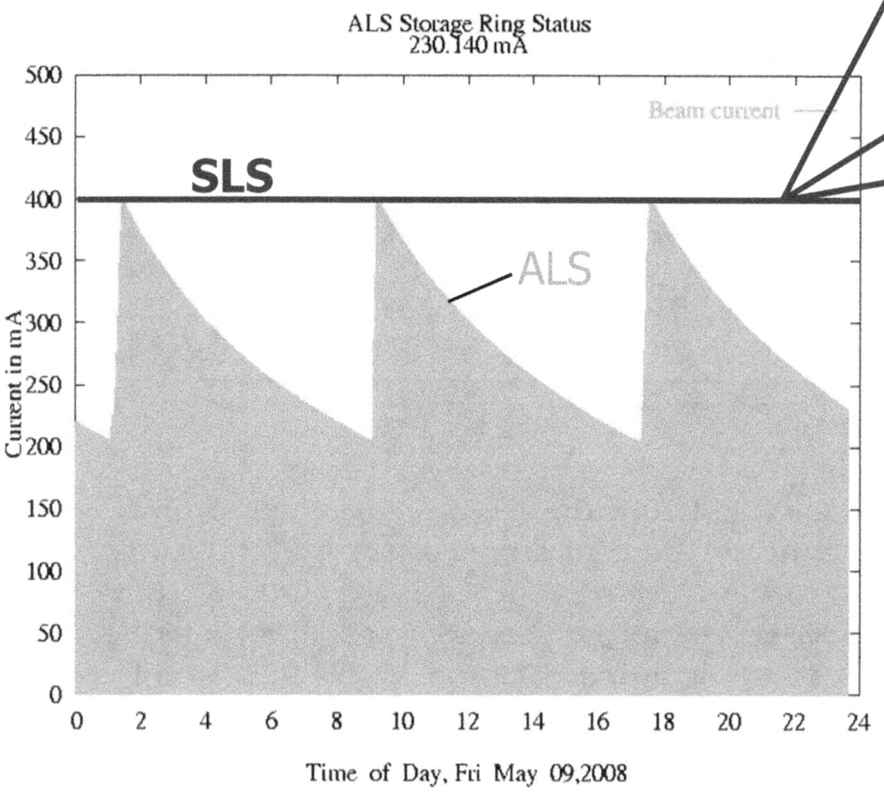

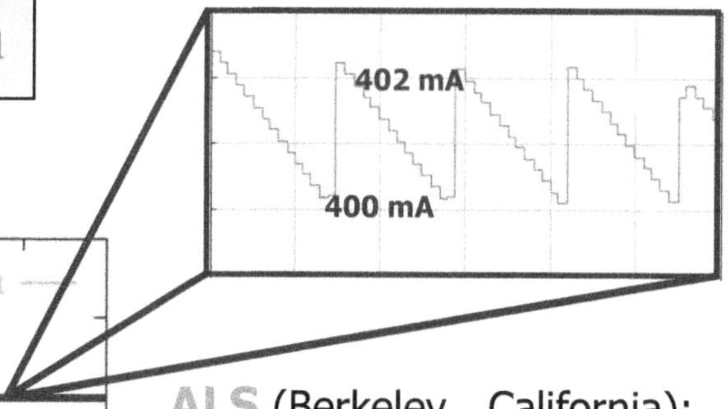

ALS (Berkeley, California):
Lifetime ~ 10 h,
Refilling every 8 h
Current: 400 => 200 mA

SLS:
Lifetime ~8 h,
not relevant !
top-up every 2.5 min.
Current: 402 => 400 mA

SLS 30° Arc

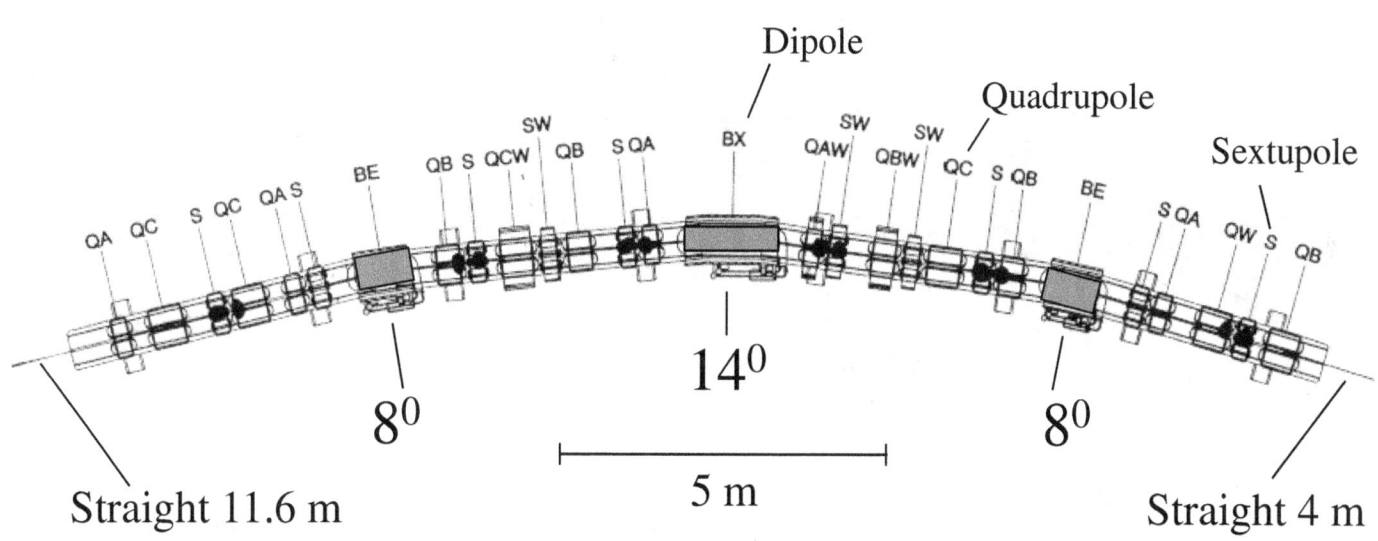

TBA-lattice

(Triple Bend Achromat)

Parameter Storage Ring

Storage ring
12 TBA: 8°/14°/8°

12 straights:
- 3 x 11 m
- 3 x 7 m
- 6 x 4 m

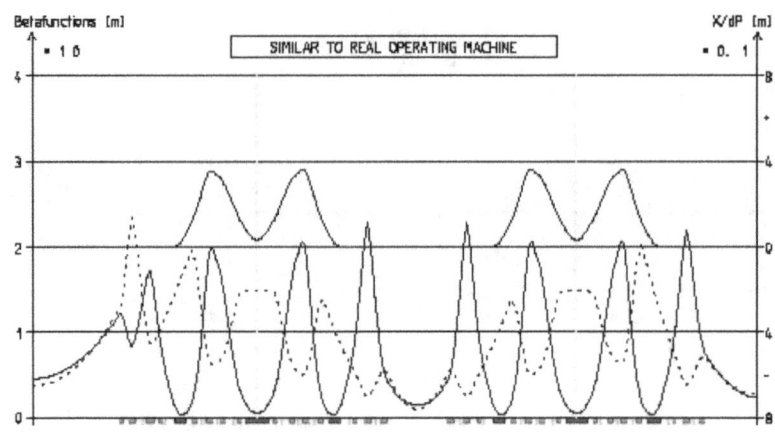

SIMILAR TO REAL OPERATING MACHINE

Energy	2.4 GeV	Emittance	5 nm rad
Circumference	288m	Bunch Length (rms)	3.5 mm
Current	400 mA	Tunes	20.41 , 8.17
RF Frequency	500 MHz	Energy Spread	9×10^{-4}
peak RF Voltage	2.6 MV	Damping Times	9, 9, 4.5 ms
Radiation loss/turn	500 keV	Momentum Compact.	0.6×10^{-3}

Superbend

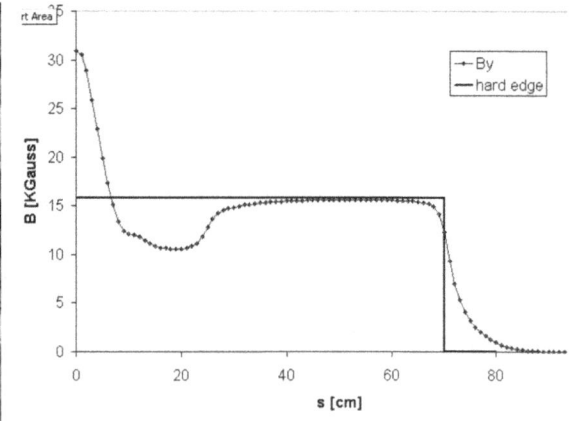

bending angle 14^0

center cone with 3 T

critical energy = 11.5 keV

end regions with 1.5 T

RF-Cavity (from Elletra)

circulating Electrons generate 200 kW of X-Rays
this power has to be refurbished by an RF-System

Cavity = Resonator
made of Copper
Frequency 500 MHz

4 Cavities in Storage Ring,
1 Cavity in Booster

600 kV Voltage
55 kW Power Loss/Cavity

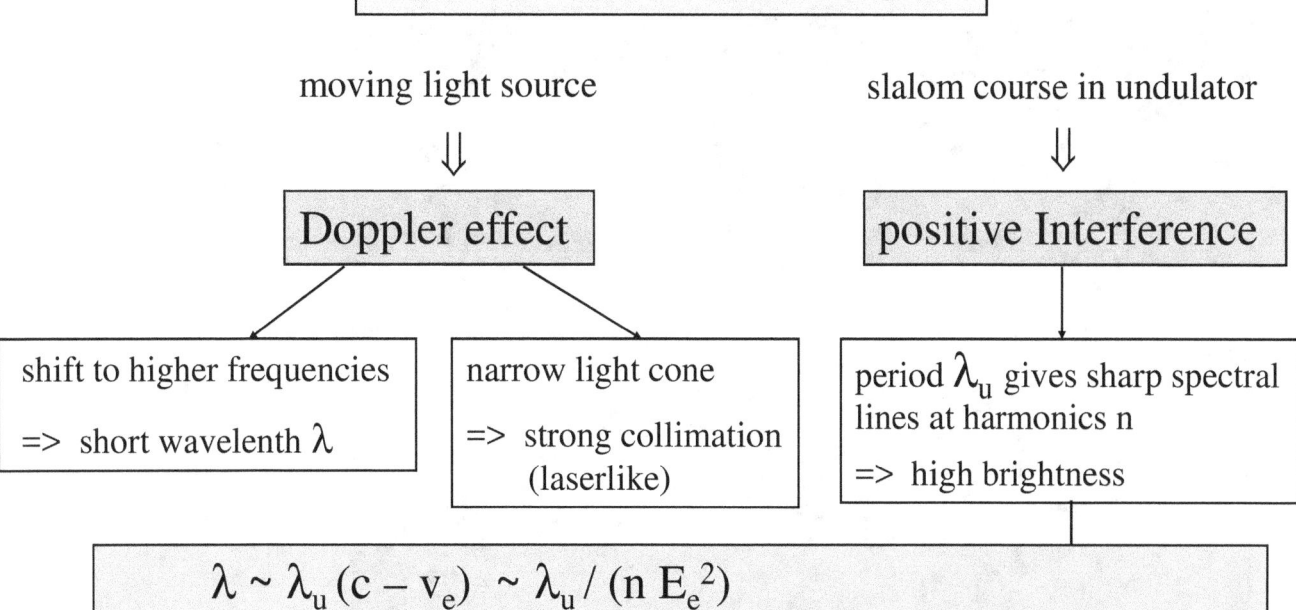

fast Electrons in magnetic field generate Synchrotron Light

moving light source ⇓ **Doppler effect**
- shift to higher frequencies => short wavelenth λ
- narrow light cone => strong collimation (laserlike)

slalom course in undulator ⇓ **positive Interference**
- period λ_u gives sharp spectral lines at harmonics n => high brightness

$$\lambda \sim \lambda_u (c - v_e) \sim \lambda_u / (n E_e^2)$$

=> short wavelength λ requires a short undulator period λ_u, a high harmonic n and a high electron energy E_e

=> large storage ring

Undulator Radiation

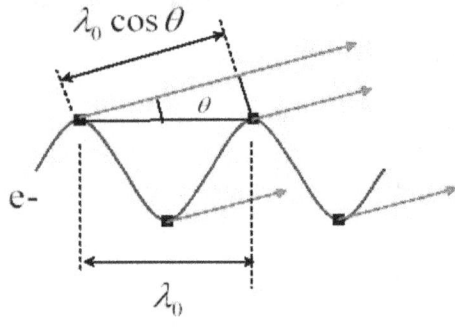

$$\lambda^* = \frac{\lambda_u}{2n}\left(1 + \frac{K^2}{2} + \gamma^2\theta^2\right)$$

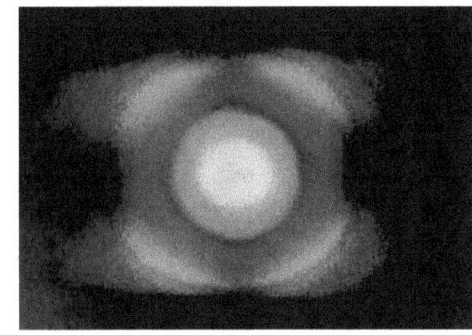

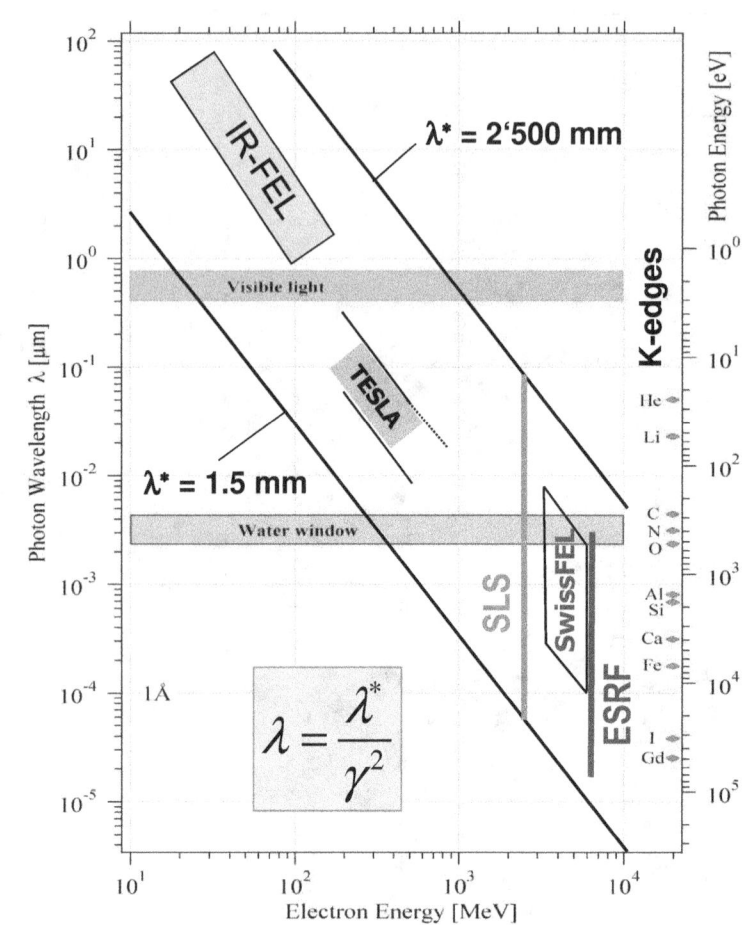

$\lambda^* = 2'500$ mm

$\lambda^* = 1.5$ mm

$$\lambda = \frac{\lambda^*}{\gamma^2}$$

Undulator UE56

Permanent Magnets

62 Periods à 56mm

helical Fields give circular and linear Polarisation

Elliptical undulator → circularly or linearly polarised light

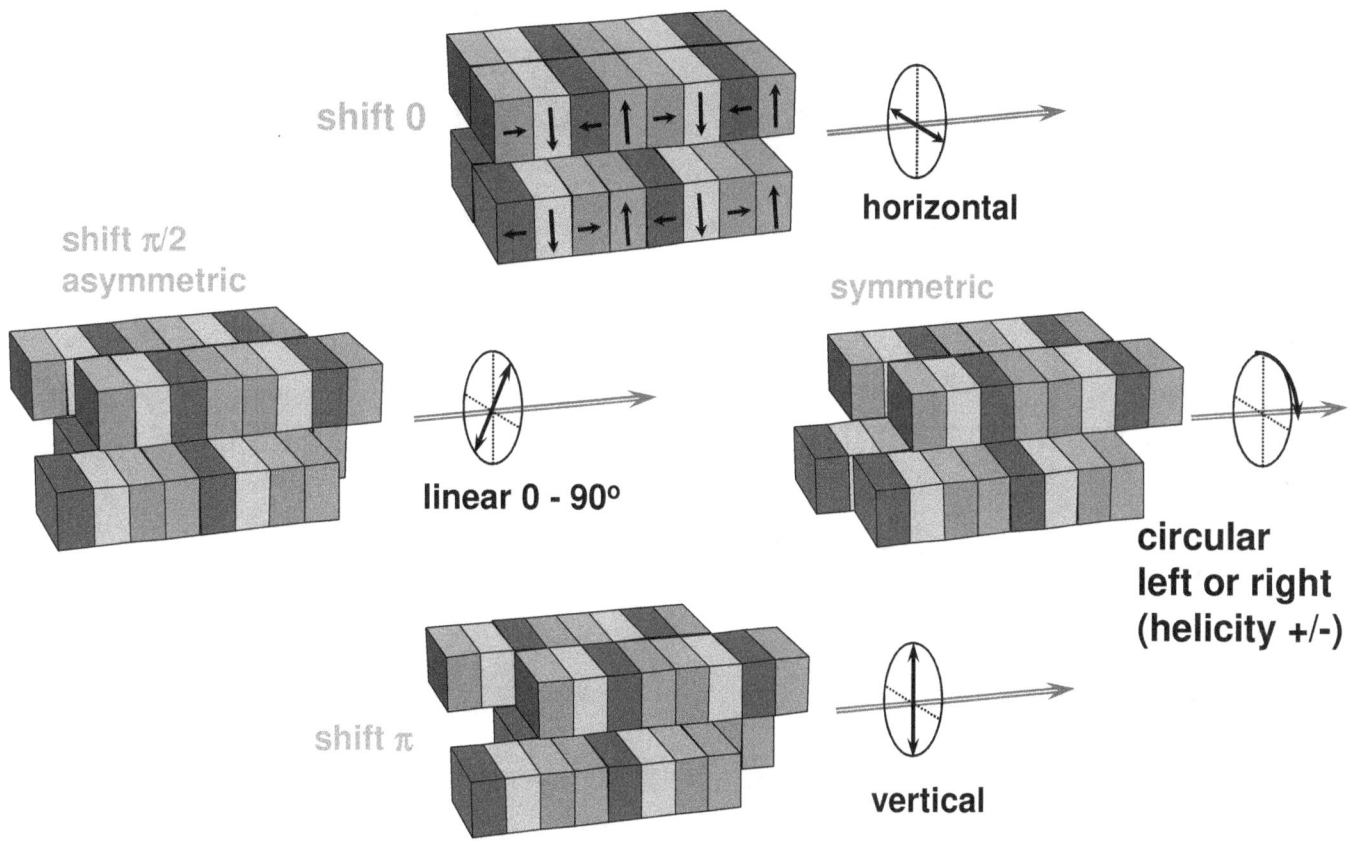

Undulator UE212, 9m long

Electromagnets

2*21 Periods à 212mm

for low energy photons (10-80eV)

helical Fields give circular and linear Polarisation

Spectrum of Undulator U19

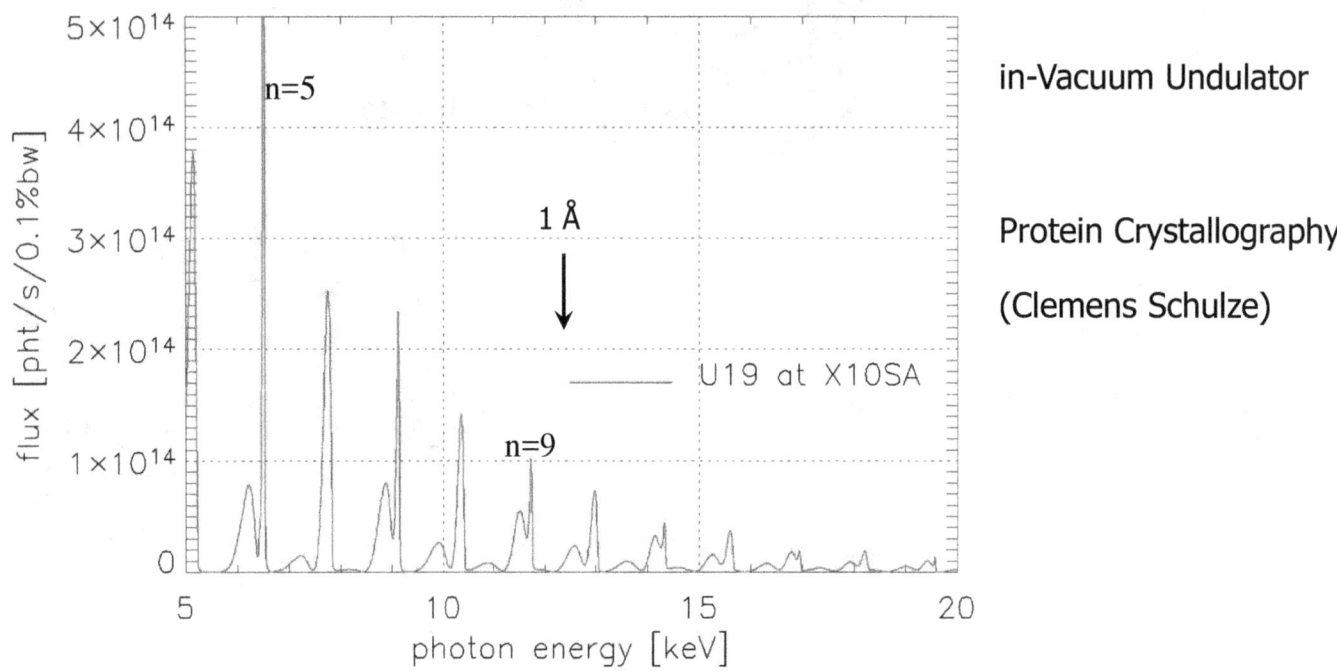

in-Vacuum Undulator

Protein Crystallography

(Clemens Schulze)

stable beams with top-up

7 days in August 2007, without interruptions!

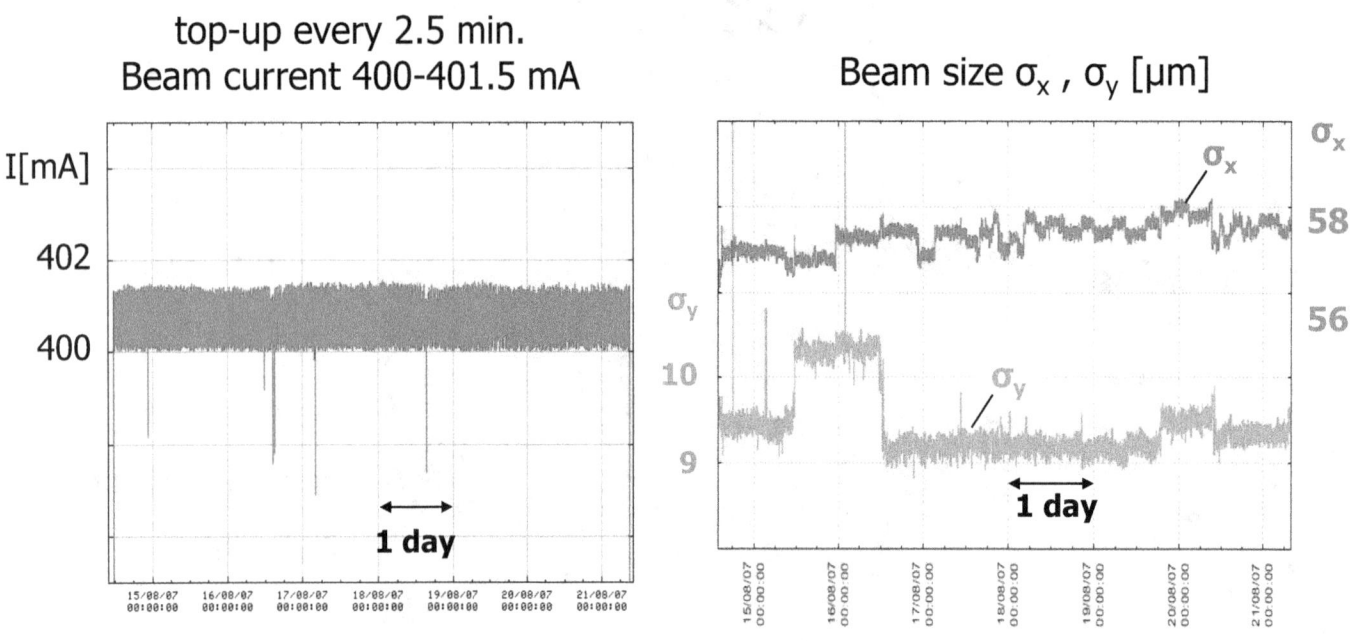

top-up every 2.5 min.
Beam current 400-401.5 mA

Beam size σ_x, σ_y [µm]

a voyage through the lung of a rat

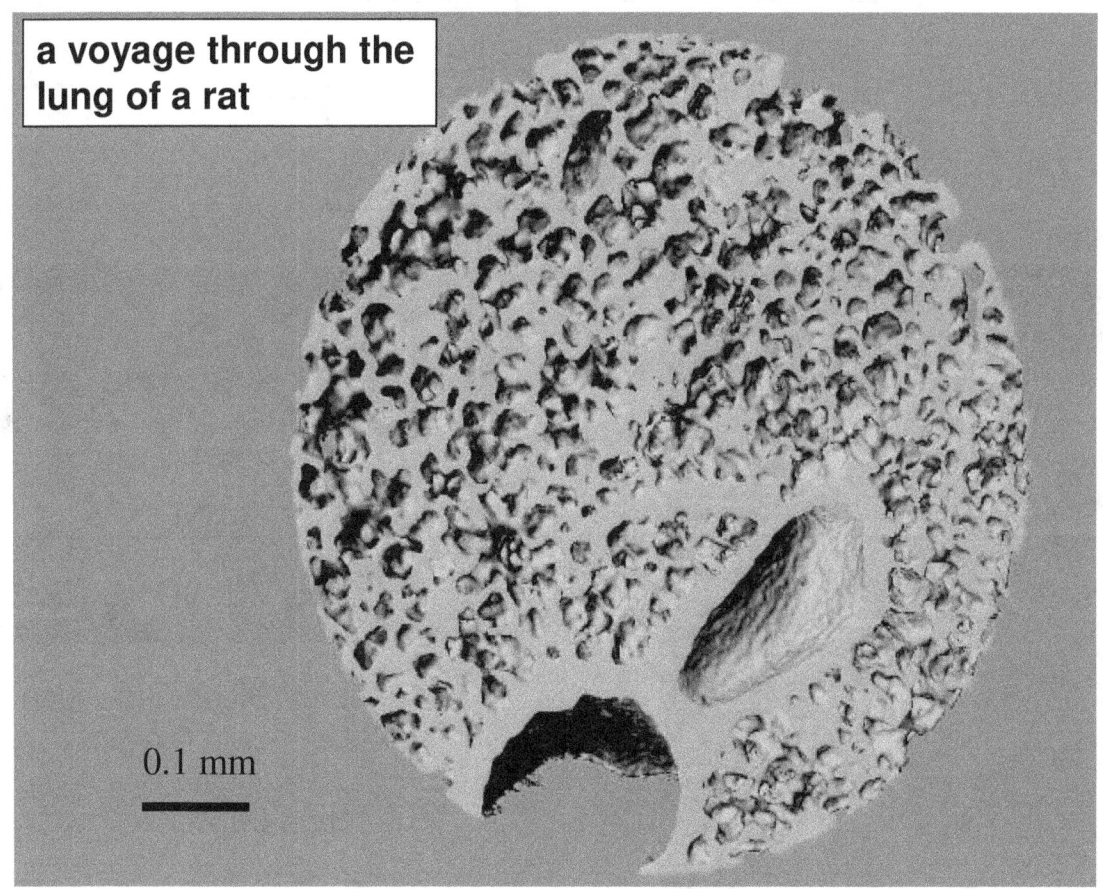

0.1 mm

u^b Prof. J. Schittny, University Bern

X-ray Tomography

Gas exchange region in the lung of a rat

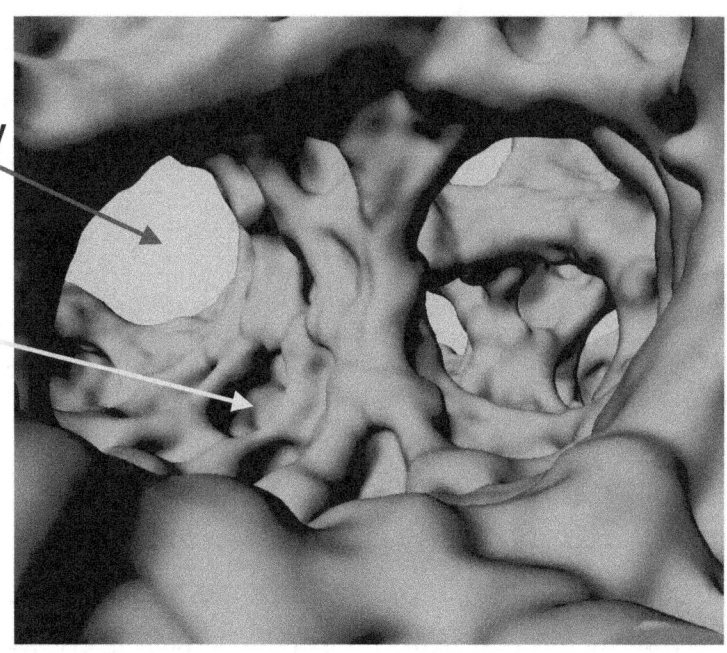

60 μm daughter airway

25 μm pouches: alveoli of lung

Prof. Schnittny (Univ. Bern) et al
Tomogram taken at beamline XO4SA-MS, M. Stampanoni et al.

Micro-Tomography
Blood Vessels in the Brain of a Mouse
(infected by Alzheimer)

uni | eth | zürich

Novartis

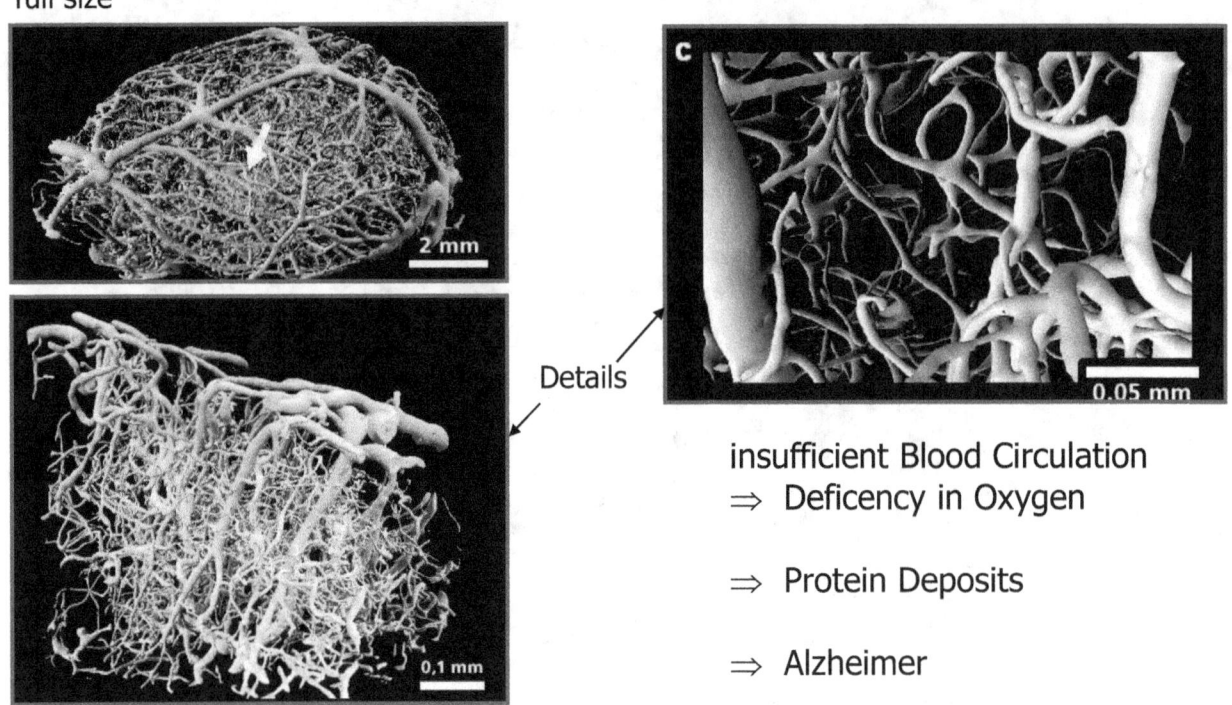

full size

Details

insufficient Blood Circulation
⇒ Deficency in Oxygen

⇒ Protein Deposits

⇒ Alzheimer

Protein Cristallography

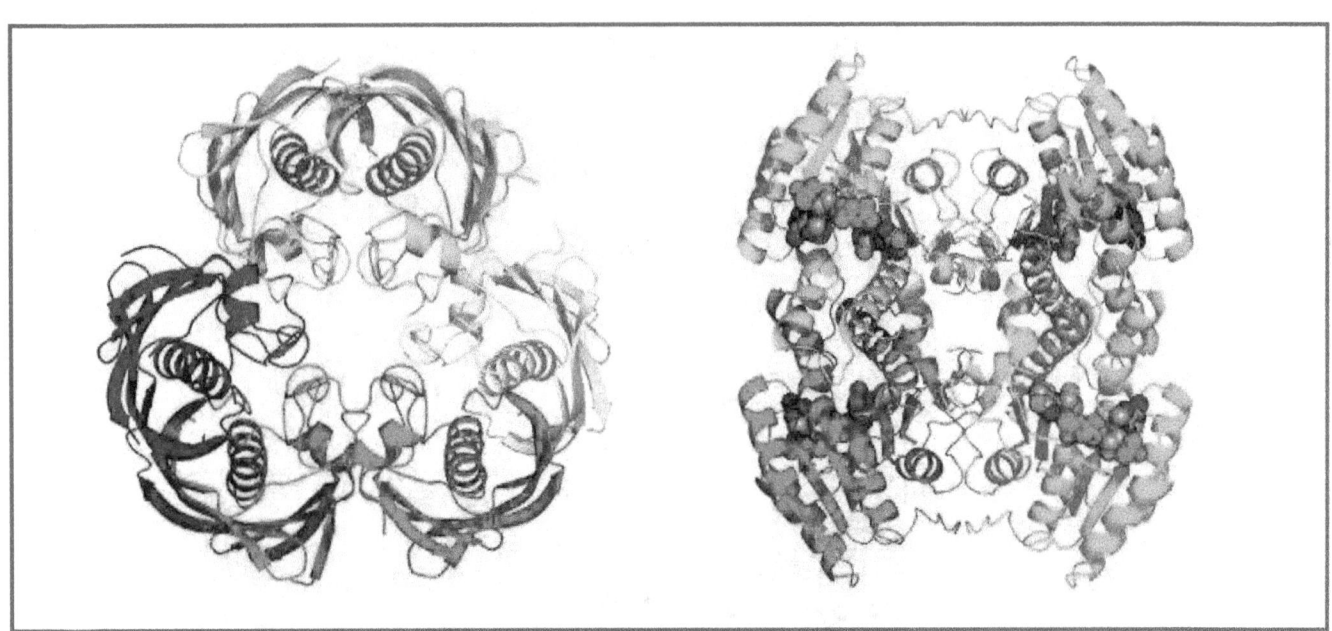

Structures of two important Enzymes of the Generator of Malarya

Growth of Bio Molecules to Crystals (size 5-50 μm)

=> Reconstruction with X-rays

magnetic Microscopy

XMCD (X-ray Magnetic Circular Dichroism)

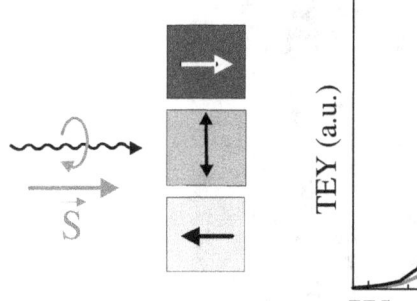
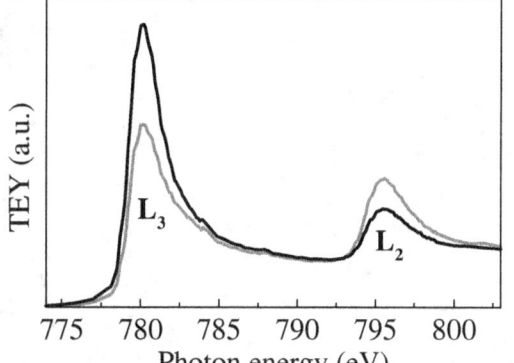

Co

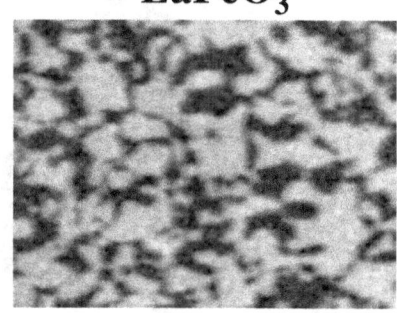

ferromagnetic

XMLD (X-ray Magnetic Linear Dichroism)

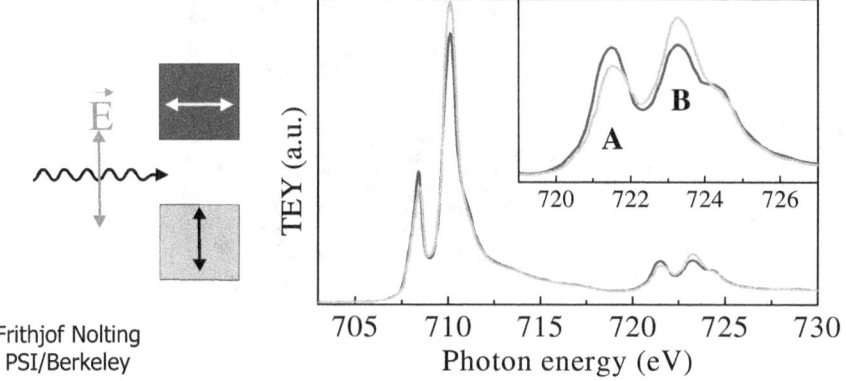

Frithjof Nolting
PSI/Berkeley

LaFeO$_3$

anti-ferromagnetic

Milestones for SLS

First Ideas	1991	Start of Building:	2.June 1998
„Giessbach-Meeting" (Users support SLS)	Oct. 1994	Building finished:	1.July 1999
ETH-council approves SLS	Sept. 1995	Beam in Linac:	23.March 2000
Parlament approves SLS	**18.June 1997**	Beam in Booster:	8.Aug. 2000
		Beam in Storage Ring	**13.Dec. 2000**
		goal of 400 mA reached	5.June 2001
		=> Begin Experiments:	2.Aug. 2001

Celebrating the success of SLS

Micha Dehler   Meinrad Eberle Director   Volker Schlott   Werner Joho   Albin Wrulich Project Leader

SLS control room

Andreas Lüdeke, Werner Joho, Michael Böge

Free Electron Laser

Combination of advantages of

Laser

- extremely short pulses
- extremely high intensity
- monochromatic light
- coherence

X-rays

- very short wavelenghts
  => details of very small structures
- transparency of materials
- adjustment of wavelength to specific atomic elements

Free Electron Laser SwissFEL

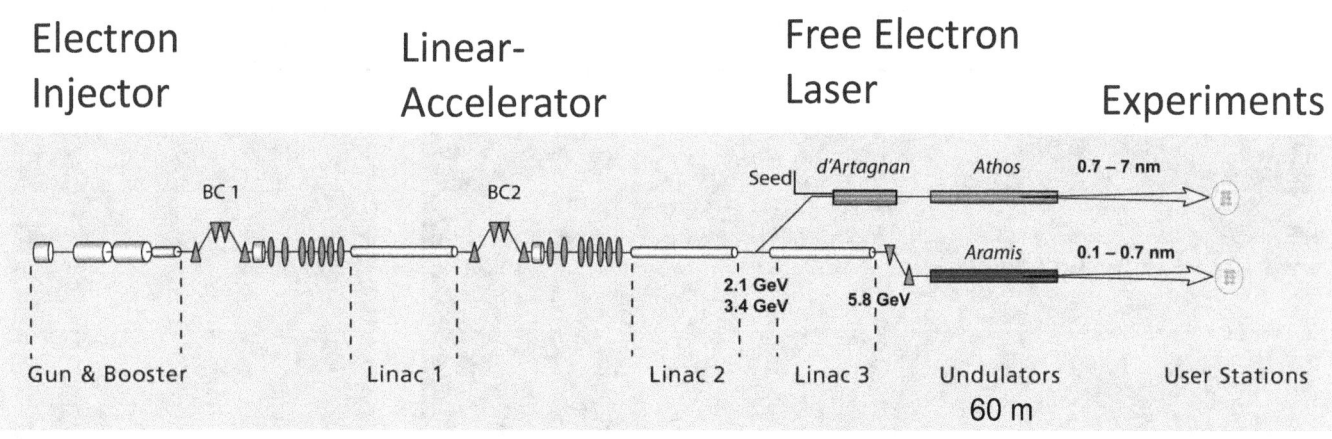

700 m

excellent beamquality of electron gun

⇒ microbunches of electrons in undulator

⇒ extremely short and intense X-ray flashes

⇒ „film of dancing molecules"

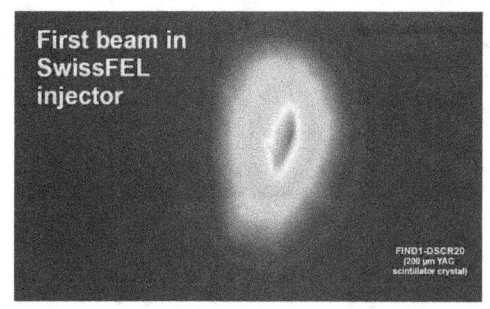

SwissFEL Wuerenlingen

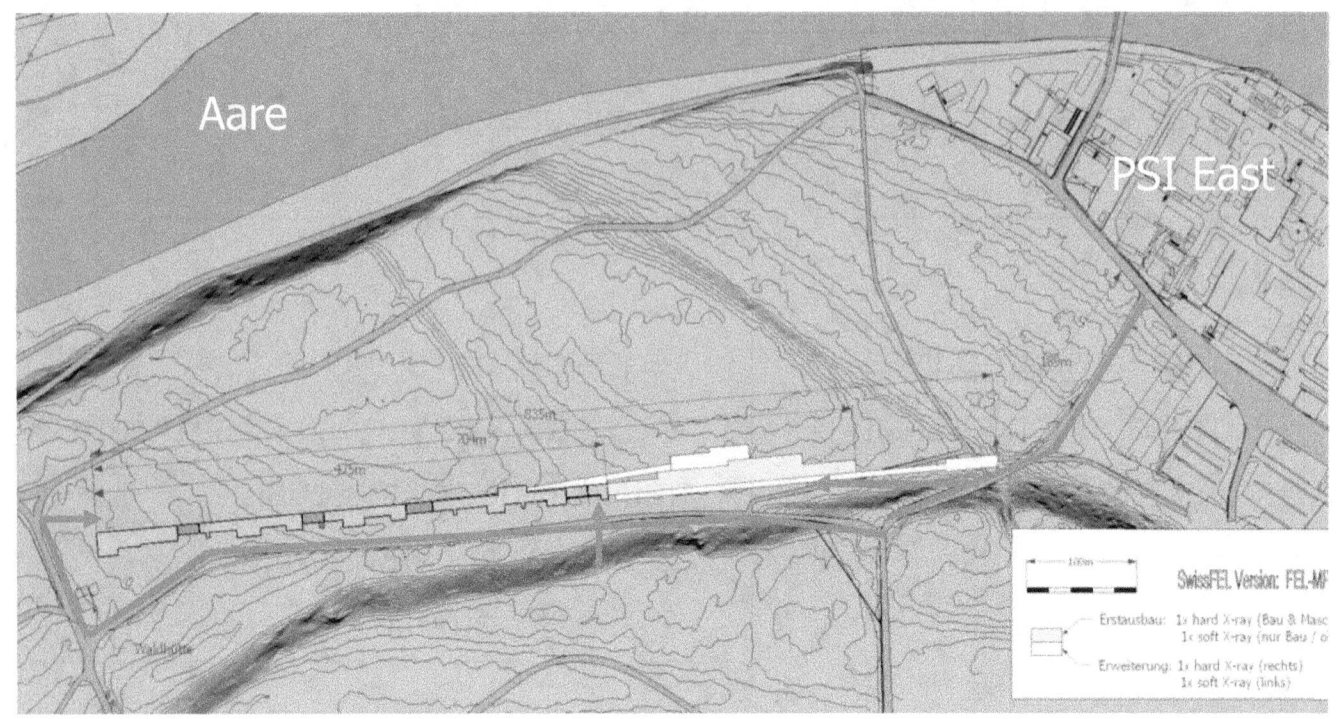

The SwissFEL during construction in the forest of Wuerenlingen

SwissFEL Electron Linear Accelerator, 6 GeV

SwissFEL Inauguration 5.December 2016

C.Quitman   A.Hürzeler   B.Moor   J.Staiblin
F.Schiesser U.Hofmann   Joel Mesot

the PSI Director explains the SwissFEL to prominent guests

SwissFEL Science
(R. Abela)

Magnetism:
materials and processes for tomorrow's information technology

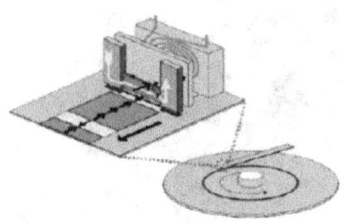

Catalysis and solution chemistry:
for a clean environment and a sustainable energy supply

Correlated electrons:
the fascination of new materials

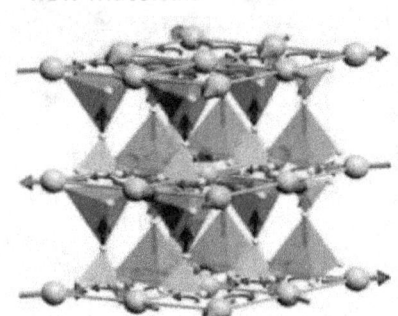

Biochemistry:
shedding light on the processes of life

Coherent diffraction:
flash photography of matter

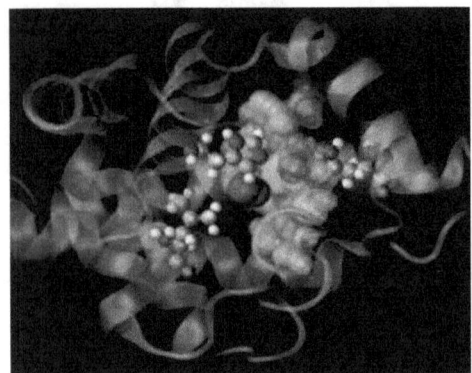

Unique Probes for Material Research

SwissFEL	=>	X-ray Flashes
SLS	=>	Synchrotron Light
SINQ	=>	Neutrons
µSR	=>	Muons

=> this combination of probes for research in

Physics, Chemistry, Biology, Material Science

at PSI

is worldwide unique !!

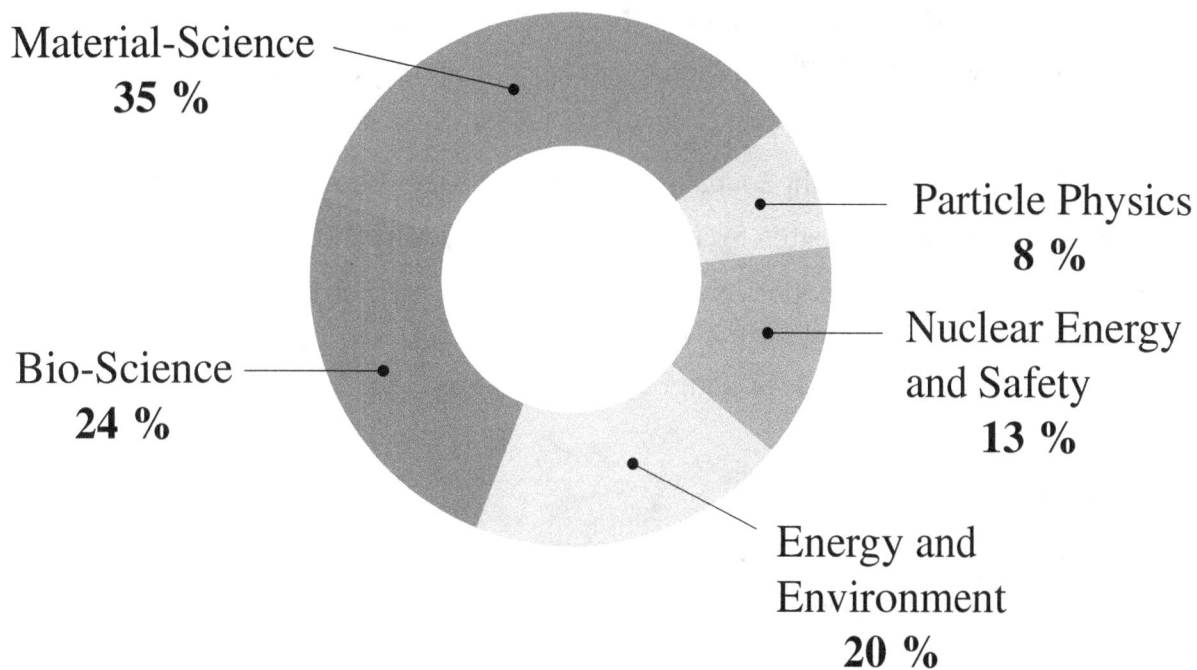

Budget for Research

- Material-Science **35 %**
- Bio-Science **24 %**
- Particle Physics **8 %**
- Nuclear Energy and Safety **13 %**
- Energy and Environment **20 %**

official retirement of the author (2003)

Werner Joho Albin Wrulich Martin Jermann Leonid Rivkin Wilfred Hirt

professional career of Werner Joho

28.8.1938 born in Baden, Switzerland

1958 – 1962   physics student at ETH Zurich

1971 PHD in physics at ETH Zurich (beam extraction from ring cyclotron)

1962 – 1990 Cyclotron group ETH => SIN/PSI , leader Beam Dynamics

1990 – 2003 Synchrotron Light Source (SLS) , Project leader Booster Synchrotron

1981-1989 Organising Committee for the International Cyclotron Conferences

1988-2000 Program and Scientific Advisory Committee for the European
            Particle Accelerator Conferences (EPAC)

> 2003 Consulting (Barcelona, Taiwan, Beijing, Vancouver, PSI)
         tour guide at PSI

external stays:

1963 CERN , Geneva, computer codes for cyclotron orbits

1963/64 Michigan State University, graduate studies

1971 – 1973 TRIUMF Vancouver, Canada, Injection Line for Cyclotron,
              lectures on thermodynamics at University of BC

1990 Berkeley, California, Advanced Light Source ALS

some personal References

W.Joho, M.Olivo, Th.Stammbach, H.Willax   "The SIN Accelerators, ..."
US Particle Accelerator Conference, 1977 Chicago, IEEE NS-24, 1618 (1977)

W.Joho "High Intensity Problems in Cyclotrons" Proc. 9th Int. Conf. on Cyclotrons,
1981, Caen, France, Les editions de physique, Paris (1981) p. 337

W.Joho, "ASTOR, concept of a combined Acceleration and Storage Ring"
US Particle Accelerator Conference, 1983 Santa Fe, IEEE NS-30, 2083 (1983)

W.Joho "Modern Trends in Cyclotrons", CERN Accelerator School, 1986 Aarhus, Denmark,
CERN 87-10, 260 (1987)

M.Böge,....W.Joho.... "The Swiss Light Source Accelerator Complex: An Overview"
6th Europ. Part. Acc. Conf. (EPAC98) Stockholm 1994, p.623

Th.Stammbach,....W.Joho,... "The feasibility of high power cyclotrons" , NIM B 113 (1996) 1-7

W.Joho, M.Munoz, A.Streun "the SLS booster synchrotron", NIM A 562 (2006) 1-11

More talks by the author are found in www.google.ch with "Werner Joho PSI"

More information on the PSI Accelerator Facilities : www.psi.ch and

Andreas Pritzker, "The Swiss Institute for Nuclear Research SIN" , Munda Verlag 2013

Key Figures 2018

PSI Funds (global budget)	280 Mio. CHF
External funding	110 Mio. CHF

Staff (heads)	2'100
externaly financed	750
PHD students	320
PSI employees teaching at ETHZ, ETHL and universities	100
Apprentices	100
External users:	2300/year
Patients treated (proton therapy)	400/year

Material Research at PSI
with X-Rays, Neutrons, Muons

SLS

Cyclotron

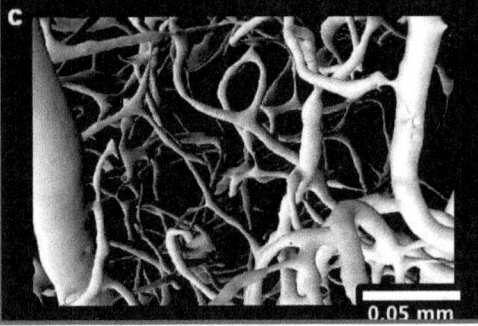

Blood Vessels

- Structure of Biomolecules for new medical drugs
- 3D-Reconstruction (Tomography) of Microstructures
- Superconductors
- Radiography of big objects (like running motor of a car) with Neutrons

Energy-Research at PSI

Alternative Energies

- Energy Storage, Power to Gas
- Fuel Cells, Hydrogen Car, efficient Batteries
- Solar-Oven, Solar Cells

Nuclear Energy

- Safety of Nuclear Power Plants
- Nuclear Waste
- Transmutation of Nuclear Waste !?

www.ingramcontent.com/pod-product-compliance
Lightning Source LLC
Chambersburg PA
CBHW081815220526
45470CB00007B/2329